D1308159

BACKYARD
BIRDS

FROM KINGFISHERS TO JAYS

This edition published in 2010 by:

CHARTWELL BOOKS, INC.
A Division of BOOK SALES, INC.
276 Fifth Avenue Suite 206
New York, New York 10001
U.S.A.

ISBN 13: 978-0-7858-2598-2
ISBN 10: 0-7858-2598-3

Printed and bound in China by Midas Printing Ltd.

Copyright © 2010 Marshall Editions

A Marshall Edition
Conceived, edited, and designed by Marshall Editions
The Old Brewery
6 Blundell Street
London N7 9BH, UK
www.quarto.com

10 9 8 7 6 5 4 3 2 1

Art Director Ivo Marloh
Editorial Elise See Tai
Production Nikki Ingram

BACKYARD BIRDS

FROM KINGFISHERS TO JAYS

DETAILED INFORMATION ON
THE TOP 100 BIRDS IN NORTH AMERICA

Editorial Consultant
Rob Hume

CHARTWELL
BOOKS, INC.

Eurasian Scops Owl
This owl has excellent camouflage and is difficult to spot.

House Martin
The House Martin is a familiar sight around human settlements.

Western Jackdaw
This bird will nest in chimney stacks and other parts of buildings.

Mallard
This adaptable bird can survive in the wild but can also be tamed in urban areas.

Ruby-topaz Hummingbird
This striking bird puts on
a dazzling display to
attract his mate.

CONTENTS

Pygmy Sunbird
This tame sunbird is easy to
spot as it perches on small
branches to find food.

INTRODUCTION

Backyards can be small nature reserves—together, they make a vast area of useful habitat for birds. They offer increasingly vital resources for wild birds while the natural riches of the countryside are under pressure. All kinds of gardens bring birds into our backyards and increasingly popular wildlife gardens are specifically designed to meet their needs. The shelter of a well-tended garden often suits birds of woodland or scrub. Lawns resemble woodland clearings. Flowerbeds and shrubberies simulate conditions in the forest shrub layer and the woodland edge.

BACKYARD HABITATS

Compared with natural habitats, gardens often have a greater density of birds, but lack nest sites and natural food supplies. In most gardens, natural holes and cavities in mature trees are few and far between and insect foods, especially caterpillars and grubs, are in short supply. Nest sites may be provided by using artificial boxes, but birds, such as titmice, chickadees, and

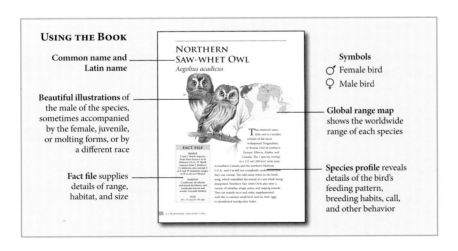

USING THE BOOK

Common name and Latin name

NORTHERN SAW-WHET OWL
Aegolius acadicus

Beautiful illustrations of the male of the species, sometimes accompanied by the female, juvenile, or molting forms, or by a different race

Fact file supplies details of range, habitat, and size

FACT FILE

RANGE
E and C North America from Nova Scotia S to N Midwest U.S.A.; W North America from S Alaska to S California; also extends S to E and W mountain ranges; in W as far as S Mexico

HABITAT
Coniferous, deciduous and mixed deciduous, and coniferous forests and woods, streamside thickets

SIZE
4¾–7½ in (17–19 cm)

This relatively tame little owl is a smaller version of the more widespread Tengmalm's or Boreal, Owl of northern Europe, Siberia, Alaska, and Canada. The 2 species overlap in a 125 ml (200 km) wide zone in southern Canada and the northern Midwest U.S.A., and it is still not completely understood how they can coexist. The odd name refers to the bird's song, which resembles the sound of a saw blade being sharpened. Northern Saw-whet Owls also utter a variety of whistles, single notes, and rasping sounds. They eat mainly mice and voles, supplemented with the occasional small bird, and lay their eggs in abandoned woodpecker holes.

6 (NORTHERN SAW-WHET OWL)

Symbols
♂ Female bird
♀ Male bird

Global range map shows the worldwide range of each species

Species profile reveals details of the bird's feeding pattern, breeding habits, call, and other behavior

Superb Sunbird
This sunbird visits forests, savanna woods, and thick bush in search of food.

house sparrows need insect food that packs a high-protein punch for their fast-growing chicks. In a forest, they glean caterpillars from foliage over a large area but in backyards, they can't find enough, and birds nesting in backyard nestboxes typically rear fewer chicks than those in natural forests. Increasingly, people supply live food, such as mealworms, but this rarely meets the demand. Normally, we feed adult birds a mixture of seeds, nuts, fatty cakes, and fruit, to give a healthy diet for small birds after they have left the nest. In winter, and especially in early spring, when natural food supplies are at their lowest ebb, this makes a huge difference to the survival of local bird populations.

HOW MANY BIRDS?

By planting carefully selected plants that offer food, shelter, and nest sites, people can meet the needs of an increasing range of wild birds. Yet our best efforts specifically attract a relatively limited range of species—in the average garden, 20 or 30 species is a good total. Many others just pass by, fly overhead, or maybe stop briefly to take advantage of what they find. A few manage to find food that we would rather they didn't have, but rarely does this get out of hand.

Rob Hume

Northern Crombec
This bird is often seen in pairs as it climbs along branches and through foliage.

Blue Jay
Cyanocitta cristata
This bird buries seeds and nuts for the winter months.

European Robin
Erithacus rubecula
This robin preys on invertebrates though it also eats seeds and fruit.

Great Spotted Woodpecker
Dendrocopos major
This woodpecker's diet varies—it feeds on insects in summer and vegetable material in winter.

Crimson Rosella
Platycercus elegans
Seeds of grasses, shrubs, and trees are the main foods for this rosella.

Birds on the Feeder

Hundreds of thousands of people put out vast amounts of bird food on bird tables and hanging feeders. Popular feeders almost everywhere offer foods, such as peanuts, maize, sunflower seeds, millet, and nyjer seeds. Woodpeckers, tits, finches, and sparrows are popular birds of hanging feeders and bird tables.

In the tropics and warmer temperate regions, birds that eat fruit and nectar come to special feeders that offer sugar solutions, or soft, sugar-rich fruits. Parrots, orioles, motmots, bulbuls, sunbirds, and hummingbirds eagerly accept such offerings and are among the most spectacularly beautiful garden birds of all.

Many birds feed on the ground or in the seclusion of low vegetation. They rarely use a hanging feeder, but feed on the fallen scraps beneath, or on foods offered in more imaginative ways, such as soft cheese scraped into tree bark. Thrushes, chats, including the European Robin, and wrens can be induced to take what we offer. Thrushes are particularly fond of fallen apples, a crucial food during the short, cold days of winter.

Ruby-throated Hummingbird
Archilochus colubris
This bird will prey on tiny flying insects and spiders.

Blue-crowned Motmot
Momotus momota
Insects and spiders are favorite foods of this bird.

race
weberi

RAINBOW LORIKEET
Trichoglossus haematodus

race
moluccanus

race
rubritorquis

race
haematodus

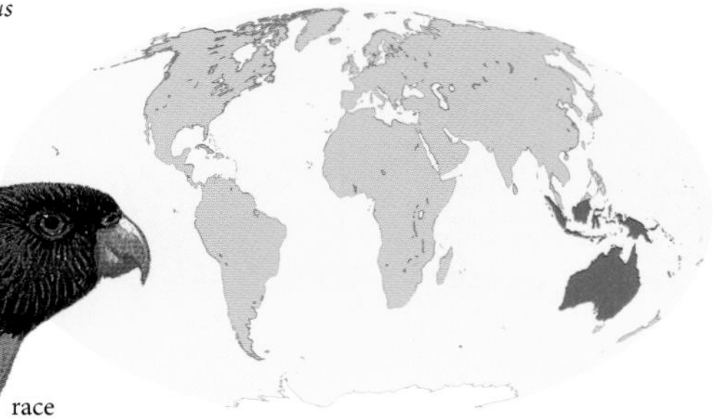

This is one of the most widespread, spectacular, and variable of the lories, with 21 races scattered over the southwest Pacific. The illustration shows *T. h. weberi* of Flores Island, *T. h. haematodus* of the southern Moluccas and western New Guinea, *T. h. moluccanus* of eastern Australia and Tasmania, and *T. h. rubritorquis* of northern Australia. Rainbow Lorikeets normally feed in the upper canopy in screeching, chattering flocks of 5–20 birds, eating fruit, insects, pollen, and nectar— they are important pollinators of coconut flowers. They spend the night in great communal roosts, sometimes with several thousand birds.

FACT FILE

RANGE
SW Pacific including
Indonesia, New Guinea,
N and E Australia

HABITAT
Rain forest, eucalypt forest,
and coconut groves up to
6,500 ft (2,000 m)

SIZE
28–32 cm (11–12 in)

ROSE-RINGED PARAKEET
Psittacula krameri

♂

B old, gregarious opportunists, Rose-ringed Parakeets will often associate in flocks of a thousand or more to feast on crops of grain or ripening fruit. They will even invade grain stores, ripping open the sacks with their hooked bills and squabbling over the spoils. During the breeding season, the flocks split up; each pair nests in a tree cavity after an elaborate courtship ritual in which the female twitters and rolls her eyes at the strutting male, rubs bills with him, and accepts gifts of food.

FACT FILE

RANGE
C Africa: E to Uganda, India, Sri Lanka; introduced to Middle and Far East, North America, England, Netherlands, Belgium, West Germany

HABITAT
Woodland, farmland

SIZE
16 in (41 cm)

CRIMSON ROSELLA
Platycercus elegans

imm

FACT FILE

RANGE
E Australia

HABITAT
Eucalypt forest, woods,
and gardens up to 6,000 ft
(1,900 m); in N restricted to
mountain forests above
1,500 ft (450 m)

SIZE
14 in (36 cm)

There are 2 very closely related rosellas—
the Yellow Rosella *P. flaveolus* and the
Adelaide Rosella *P. adelaidae*—which
are considered by some ornithologists
to belong to the same species as the
Crimson Rosella. However, their coloration
is so different that they are frequently classified as
separate species. The Crimson Rosella is the most
striking, both sexes having the same rich red and
blue plumage. In immature birds, much of the red is
replaced by green. The adults occur in pairs or small
groups and stay in the same area throughout the
year, but the immature birds are nomadic, wandering
in flocks of up to 30. They feed mainly on the seeds
of grasses, shrubs, and trees such as eucalypts and
acacias, but will also eat fruit, nectar, and insects.
They nest in tree cavities.

RUBY-TOPAZ HUMMINGBIRD

Chrysolampis mosquitus

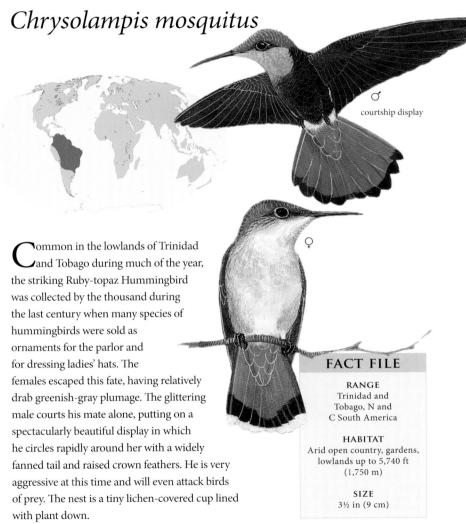

♂
courtship display

♀

Common in the lowlands of Trinidad and Tobago during much of the year, the striking Ruby-topaz Hummingbird was collected by the thousand during the last century when many species of hummingbirds were sold as ornaments for the parlor and for dressing ladies' hats. The females escaped this fate, having relatively drab greenish-gray plumage. The glittering male courts his mate alone, putting on a spectacularly beautiful display in which he circles rapidly around her with a widely fanned tail and raised crown feathers. He is very aggressive at this time and will even attack birds of prey. The nest is a tiny lichen-covered cup lined with plant down.

FACT FILE

RANGE
Trinidad and
Tobago, N and
C South America

HABITAT
Arid open country, gardens,
lowlands up to 5,740 ft
(1,750 m)

SIZE
3½ in (9 cm)

RUBY-THROATED HUMMINGBIRD
Archilochus colubris

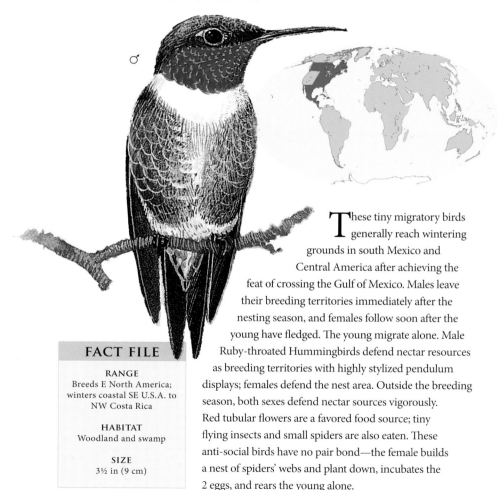

♂

These tiny migratory birds generally reach wintering grounds in south Mexico and Central America after achieving the feat of crossing the Gulf of Mexico. Males leave their breeding territories immediately after the nesting season, and females follow soon after the young have fledged. The young migrate alone. Male Ruby-throated Hummingbirds defend nectar resources as breeding territories with highly stylized pendulum displays; females defend the nest area. Outside the breeding season, both sexes defend nectar sources vigorously. Red tubular flowers are a favored food source; tiny flying insects and small spiders are also eaten. These anti-social birds have no pair bond—the female builds a nest of spiders' webs and plant down, incubates the 2 eggs, and rears the young alone.

FACT FILE

RANGE
Breeds E North America; winters coastal SE U.S.A. to NW Costa Rica

HABITAT
Woodland and swamp

SIZE
3½ in (9 cm)

GREAT SPOTTED WOODPECKER
Dendrocopos major

race *pinetorum*

♀

juv

♂

The Great Spotted Woodpecker is widely distributed and shows considerable geographical variation. Most Great Spotted Woodpeckers are sedentary, but some northern populations may migrate south in winter to find food. Although rather secretive at times, these birds often betray their presence with a loud *tchick* call, and on warm spring days, the woods echo with the sound of their courtship and territorial drumming. Their diet varies seasonally to include an abundance of insects in summer and vegetable material, including seeds, in winter. The birds use branches or stumps of trees as anvils to break open seeds in winter, and can wedge pinecones into crevices in order to peck out the kernels with their bills. Occasionally, during the breeding season, these woodpeckers hack their way into nest-boxes or natural cavities and kill and eat the chicks of other hole-nesting birds, especially tits.

FACT FILE

RANGE
N Eurasia to Middle East
and N Africa

HABITAT
Deciduous and coniferous
forest from Arctic taiga to
Mediterranean scrub

SIZE
8½–9 in (22–23 cm)

HAIRY WOODPECKER
Picoides villosus

race *septentrionalis*

♂

race *sanctorum*

♂

FACT FILE

RANGE
Alaska and Canada,
S to Central America,
W Panama and Bahamas

HABITAT
Coniferous and
deciduous forest

SIZE
7–9½ in (18–24 cm)

The Hairy Woodpecker is, in many respects, a New World counterpart of the Great Spotted Woodpecker. It is one of the most widely distributed and geographically variable of all woodpeckers. For instance, the race *P. v. septentrionalis*, from Alaska, Canada, and parts of the central and western U.S.A., has striking black and white plumage and may weigh up to 5 oz (135 g), whereas the race *P. v. sanctorum*, from southern Mexico and Central America, has black, white, and chocolate-brown coloring and is much smaller, weighing only about 1½ oz (38 g). Northern populations may migrate south for food in winter, but most birds are sedentary. Hairy Woodpeckers forage on trunks and major limbs of trees for surface and subsurface insects, berries, and seeds. Their calls include a loud *peek* location call and a rattling call.

RED-HEADED WOODPECKER

Melanerpes erythrocephalus

juv

Both males and females have the completely red head
that the juveniles lack until the spring following their
year of hatching. They are highly aggressive birds whose red
heads may help them when they are usurping nest-holes
from other species of woodpeckers (typically the Red-bellied
Woodpecker *M. carolinus*)—the striking head pattern seems
to scare off the other birds. Unusually for woodpeckers, the
species catches flying insects in midair and swoops down
to capture grasshoppers and other insects from the ground,
rather than gleaning them from tree surfaces. The latter habit
often exposes them to traffic dangers and many are killed
on the roads. In winter, migrant Red-headed Woodpeckers
congregate in oak woodlands, where they feed on acorns.
Nest sites, typically in isolated trees, but sometimes in utility
poles, are often taken from other species. New holes may
be excavated in soft wood. Clutches generally consist of 3–5
eggs. Both parents incubate the eggs and feed the young.

FACT FILE

RANGE
E U.S.A. and S Canada

HABITAT
Open woods, parks,
scattered trees

SIZE
7½ in (19 cm)

Northern Flicker
Colaptes auratus

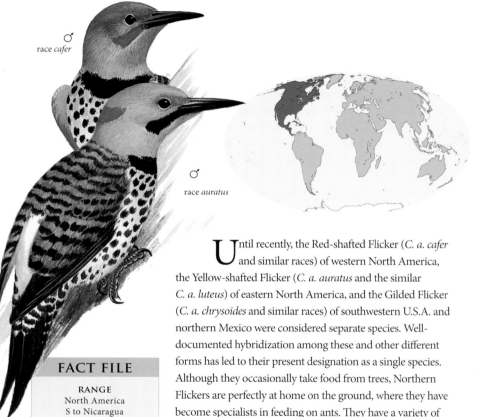

♂
race *cafer*

♂
race *auratus*

FACT FILE

RANGE
North America
S to Nicaragua

HABITAT
Open forest, savanna,
open areas with widely
scattered trees

SIZE
10–14 in (25.5–36 cm)

Until recently, the Red-shafted Flicker (*C. a. cafer* and similar races) of western North America, the Yellow-shafted Flicker (*C. a. auratus* and the similar *C. a. luteus*) of eastern North America, and the Gilded Flicker (*C. a. chrysoides* and similar races) of southwestern U.S.A. and northern Mexico were considered separate species. Well-documented hybridization among these and other different forms has led to their present designation as a single species. Although they occasionally take food from trees, Northern Flickers are perfectly at home on the ground, where they have become specialists in feeding on ants. They have a variety of calls and both sexes produce the characteristic woodpecker drumming sound with their bills. In the absence of suitable trees for drumming, they often drum on buildings, television antennas, and other artificial structures. Although they can excavate nest-holes in soft wood or cacti, they frequently use natural cavities or renovate old nests.

BLUE-CROWNED MOTMOT

Momotus momota

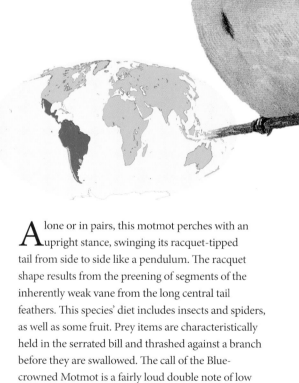

A lone or in pairs, this motmot perches with an upright stance, swinging its racquet-tipped tail from side to side like a pendulum. The racquet shape results from the preening of segments of the inherently weak vane from the long central tail feathers. This species' diet includes insects and spiders, as well as some fruit. Prey items are characteristically held in the serrated bill and thrashed against a branch before they are swallowed. The call of the Blue-crowned Motmot is a fairly loud double note of low frequency, sometimes uttered in a long sequence: *hudu … hudu, du, du, du.*

FACT FILE

RANGE
E Mexico to NW Peru, NW Argentina, and SE Brazil; Trinidad and Tobago

HABITAT
Rain forest, deciduous forest, coffee and cocoa plantations, semi-open habitats

SIZE
41 cm (16 in)

EUROPEAN ROBIN
Erithacus rubecula

FACT FILE

RANGE
Europe and N Africa
E to W Siberia and N Iran

HABITAT
Woodland and forest; also
gardens, parks in Britain

SIZE
5½ in (14 cm)

The European Robin has an upright, bold stance on the ground, regularly flicking its wings and tail. It is commonly seen as it flits from one low perch to another and the red breast figures prominently. Continental European races are shy, but in Britain, where this species is the national bird, many can be seen around human habitations. Most populations are resident and their songs can be heard for much of the year. The European Robin is largely monogamous and both sexes defend winter territories. Although they are very aggressive birds, they rarely make physical contact during fights. When they do, one may peck the other to death. Invertebrates form the basic diet of European Robins, although the birds will eat seeds and fruit during cold winters. Nesting takes place in hollows in banks, holes in trees and walls, and in an extraordinary variety of artificial items including discarded cans or kettles. The nest is a cup of moss, feathers, and sometimes plastic strips, built on a base of leaves.

BLACKBIRD
Turdus merula

♀

♂

The Blackbird is a familiar species throughout western Europe, where its melodic, fluty song is a common sound over a wide range of habitats. From its original woodland haunts, the bird has spread into gardens and even into the centers of towns and cities. Today, gardens harbor the densest populations, followed by parks and farmland. Densities are lowest deep in woodlands, where the birds are often much more shy. Blackbirds prise earthworms from the soil all year round and, by avidly turning over leaf litter, they catch a variety of insects and other invertebrates. They have a mixed diet and at certain times of the year, the fruits of hawthorn, holly, elder, and yew provide an important source of food. Their nest is a substantial cup of grasses or roots, cemented together with wet leaves or mud. The usual clutch contains 3–5 pale greenish eggs with red-brown speckles.

FACT FILE

RANGE
Breeds NW Africa, Europe E to India, S China; N and some E populations winter S to Egypt, SW Asia, Southeast Asia; introduced to Australia, New Zealand

HABITAT
Diverse, including forest, farmland, moors, scrub, gardens, parks, inner city

SIZE
9½–10 in (24–25 cm)

AMERICAN ROBIN
Turdus migratorius

juv

FACT FILE

RANGE
North America

HABITAT
Forest borders, woodland,
parks, lawns, suburbs

SIZE
9–11 in (23–28 cm)

Like the much smaller European Robin, this species is also known colloquially as the Robin Redbreast. It is the largest of the North American thrushes and is a very common and well-known bird. It occurs anywhere with cover, from suburbs to high mountainsides. American Robins usually feed on grassy ground, locating them by sight. Their diet also includes insects, snails, and much fruit, especially in winter. In the southern portion of its range, the American Robin has both resident and migratory populations—some males move north in February, followed by females in April. The song is musical, with short, rising and falling phrases. Both the male and female build a bulky nest of twigs, grass, and mud, raising 2 or 3 broods each year. The male cares for the young of the first brood, while the female incubates the second clutch.

BLACK-CAPPED CHICKADEE
Poecile atricapillus

This is the most widely distributed tit in North America, and one of 6 species known as chickadees from their calls. It is a familiar garden visitor in winter, when small parties of 6 or so birds regularly descend on feeding stations, but in spring these parties split up into nesting pairs, which spend the summer feeding in the forest. The pair excavate their own nest-hole in the soft wood of a dead tree. They usually ignore any vacant holes, but will sometimes use a nest-box if it is previously stuffed with debris so that they have to dig their way in.

FACT FILE

RANGE
Alaska, Canada,
S to C U.S.A.

HABITAT
Coniferous and
broad-leaved forest

SIZE
5 in (13 cm)

COAL TIT
Periparus ater

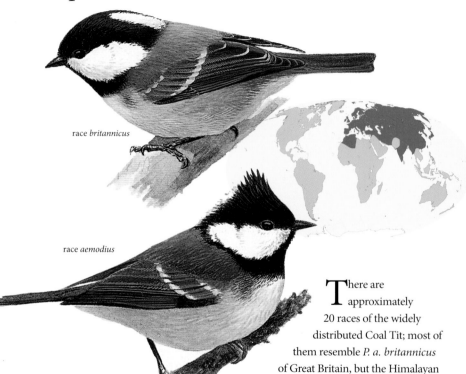

race *britannicus*

race *aemodius*

There are approximately 20 races of the widely distributed Coal Tit; most of them resemble *P. a. britannicus* of Great Britain, but the Himalayan race *P. a. aemodius* has a striking black crest. In general, they favor conifer forests although some, such as those in Ireland, are found mainly in broad-leaved woodlands. Outside of the breeding season, they often feed in mixed flocks, searching through the tree canopy for insects and seeds. Coal Tits do not excavate their own nest-holes—they prefer to nest in vacant holes in trees, but they will sometimes use abandoned rodent burrows in the ground.

FACT FILE

RANGE
Eurasia from Britain to Japan, N Africa

HABITAT
Forests, mainly coniferous

SIZE
4½ in (11.5 cm)

GREAT TIT
Parus major

race *major*

race *cinereus*

The Great Tit is a very widespread, adaptable bird, and there are about 30 distinct races. Many of these are yellow-bellied, green-backed birds like *P. m. major* of Europe and central Asia, but several eastern races, including *P. m. cinereus* of Java and the Lesser Sunda Islands, have a gray back and a whitish belly. Green/yellow forms and gray/whitish ones occur together in the eastern U.S.S.R., but do not seem to interbreed. Acrobatic and vocal, the Great Tit has a wide vocabulary of metallic call notes. It is a common visitor to gardens, where it readily breeds in nest-boxes as well as in such unlikely sites as drainpipes and mailboxes.

FACT FILE

RANGE
Eurasia, from Britain through U.S.S.R. and S Asia to Japan

HABITAT
Forests, mountain scrub, parks and gardens, mangroves

SIZE
5½ in (14 cm)

BLUE TIT
Cyanistes caeruleus

FACT FILE

RANGE
Europe E to Volga, Asia
Minor, N Africa

HABITAT
Woods, parks, gardens

SIZE
4 in (11 cm)

These lively birds are popular visitors to suburban gardens in winter, giving virtuoso displays of agility as they cling to suspended coconuts or whisk tasty morsels away from beneath the bills of larger, slower birds. In their natural woodland habitat they often feed in loose mixed flocks with other tits, such as the Great Tit and Coal Tit, working steadily from tree to tree. An eager user of garden nest-boxes, the Blue Tit lays the largest clutch of any bird in the world (excluding some ducks and gamebirds, which do not feed their young). In good habitats, 11 eggs are common, while clutches of as many as 14–15 are by no means rare.

LONG-TAILED TIT
Aegithalos caudatus

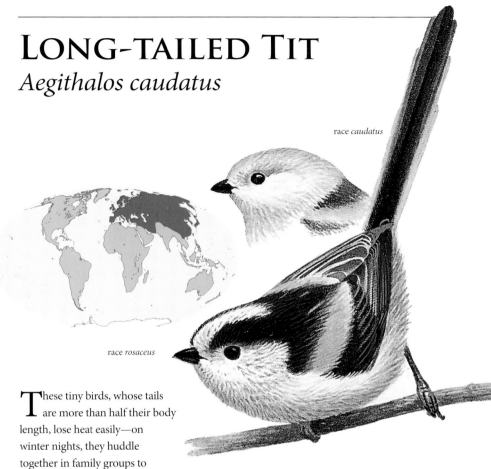

race *caudatus*

race *rosaceus*

These tiny birds, whose tails are more than half their body length, lose heat easily—on winter nights, they huddle together in family groups to keep warm. When day breaks, the groups set off to forage together through the treetops, often in mixed flocks with other small birds. In spring, they pair up to build beautifully woven, domed nests of lichens, spiders' silk, and feathers. If a pair loses its brood, they will often help to feed the broods of relatives. There are some 20 races. The British race *A. c. rosaceus* has a distinctive broad black stripe above the eye, in contrast to the white-headed northern race *A. c. caudatus.*

FACT FILE

RANGE
Continuous from Europe
through C Asia to Japan

HABITAT
Woods, woodland
edge, scrub

SIZE
5½ in (14 cm)

TUFTED TITMOUSE

Baeolophus bicolor

race *atricristatus*

race *bicolor*

FACT FILE

RANGE
E North America, S to Texas
and currently extending
N to Ontario

HABITAT
Broad-leaved forest,
open woods

SIZE
6½ in (17 cm)

Many tits owe their winter survival to bird lovers who put food out for them. The Tufted Titmouse is no exception, and it is likely that this is the reason why it is spreading farther north. Even without this help, it is well equipped to gather winter food, for its stout bill enables it to hammer open nuts that would be too tough for most other tits to crack. The race *B. b. bicolor* is from eastern, central, and southeastern U.S.A., and is typical of the species. At the other end of its range, in Texas, the bird has a longer, black crest. Once known as the Black-crested Titmouse, this is now generally thought to be a race, *B. b. atricristatus*, of the Tufted Titmouse.

EURASIAN NUTHATCH
Sitta europaea

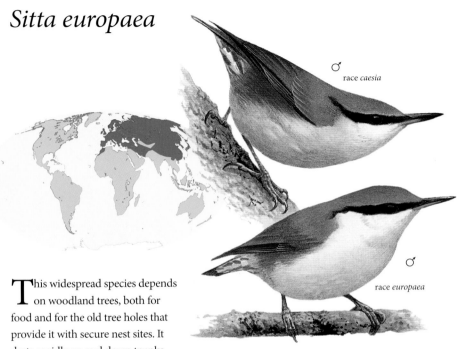

♂ race *caesia*

♂ race *europaea*

This widespread species depends on woodland trees, both for food and for the old tree holes that provide it with secure nest sites. It darts rapidly up and down trunks and branches, searching for invertebrates, seeds, and nuts. It often wedges large items in crevices and hammers them vigorously with its sharp bill to break them open. The race *S. e. europaea* from Scandinavia, eastern Europe, and central Russia has very pale underparts, in contrast to races, such as *S. e. caesia* of western Europe, including England and Wales. The 2 forms have met up after a long period of isolation and are intermingling. The nest usually lies in an abandoned woodpecker hole toward the top of a tall tree and is freshly floored with flakes of pine bark and lichen. A "plaster" of dried clay and mud is used to reduce the size of the entrance hole.

FACT FILE

RANGE
W Europe E to Japan and Kamchatka (U.S.S.R.)

HABITAT
Woodland with large trees

SIZE
4–5 in (11–13 cm)

Rufous-sided Towhee
Pipilo erythrophthalmus

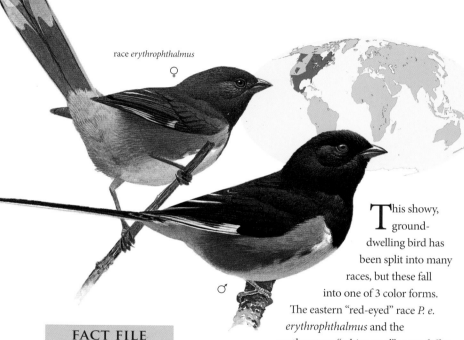

race *erythrophthalmus*

♀

♂

This showy, ground-dwelling bird has been split into many races, but these fall into one of 3 color forms. The eastern "red-eyed" race *P. e. erythrophthalmus* and the southeastern "white-eyed" races differ only in eye color; the western form, including such races as *P. e. oregonus*, differs in its white wing bars and white-spotted scapulars. Rufous-sided Towhees enjoy a varied diet, from weed seeds and wild berries to insects. They commonly scratch through leaf litter and readily visit bird feeders for suet and seeds. Their stout cup nests are placed on the ground, hidden in dense undergrowth or brush, or low in dense thickets and vine tangles.

FACT FILE

RANGE
SW Canada, U.S.A.
S to Baja California

HABITAT
Dense undergrowth,
streamside thickets, open
woodland forest edge

SIZE
8½ in (22 cm)

DARK-EYED JUNCO
Junco hyemalis

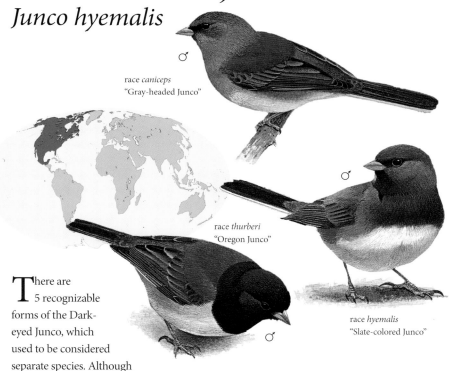

race *caniceps*
"Gray-headed Junco"

♂

race *thurberi*
"Oregon Junco"

♂

race *hyemalis*
"Slate-colored Junco"

♂

There are
5 recognizable
forms of the Dark-
eyed Junco, which
used to be considered
separate species. Although
they vary considerably in plumage, they all have white
outer tail feathers, a white belly, and dark eyes. The
different forms interbreed freely where their ranges meet.
Of those illustrated, *J. h. hyemalis* is common in the boreal
forests of northern Canada and northern central U.S.A.;
J. h. thurberi breeds in coniferous forests of the far west;
and *J. h. caniceps* occurs in the southern Rockies west to
eastern California. All forms of the Dark-eyed Junco sing
a musical trill, and utter a rapid twitter in flight. Although
primarily seed-eaters, they feed on many insects during
the nesting season.

FACT FILE

RANGE
Breeds Canada, N and
C U.S.A.; N populations
migrate S as far as Mexico

HABITAT
Forests of conifers, birch,
aspen; various on migration
and in winter

SIZE
6 in (16 cm)

NORTHERN CARDINAL
Cardinalis cardinalis

FACT FILE

RANGE
SE Canada, E, C, and SW
U.S.A., Mexico to Belize;
introduced to Hawaii

HABITAT
Woodland edges, swamps,
thickets, suburban gardens

SIZE
8½ in (22 cm)

Among the most regular and popular of birds at feeding stations, Northern Cardinals often visit in pairs or family groups to take sunflower and safflower seeds. The wild food they eat depends on upon availability, but includes insects, fruit, seeds, blossoms, buds, and the residues that collect in holes drilled by sapsuckers. The loud, whistling song is uttered by both males and females almost all year round. A prolonged nesting season for this species may result in up to 4 broods.

ROSE-BREASTED GROSBEAK
Pheucticus ludovicianus

This thick-billed bird has a special fondness for blossoms and buds, although it also eats beetles, grasshoppers, cankerworms, seeds, and some fruit and grain. It often visits feeding stations, especially during migration. The male often delivers his liquid melodies in flight, while pursuing a female. The drabber female also sings, producing a shorter, softer tune. Mates sometimes touch bills in courtship. Both share the nesting duties—the male later caring for the fledglings while the female re-nests. The male molts into a duller, brown-tipped winter plumage before migrating.

FACT FILE

RANGE
SC Canada, E U.S.A.;
winters S to Mexico
and N South America

HABITAT
Secondary growth woodland,
trees along watercourses

SIZE
8 in (20 cm)

SCARLET TANAGER
Piranga olivacea

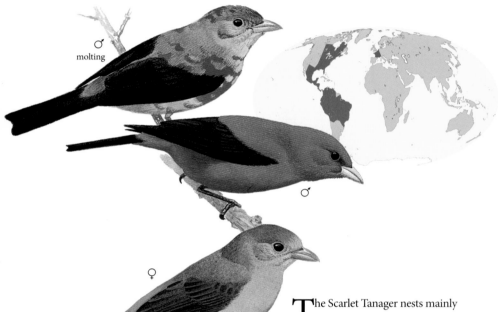

♂ molting

♂

♀

The Scarlet Tanager nests mainly in deciduous forests in eastern U.S.A., where it feeds on bees, wasps, beetles, moths and their larvae, supplemented by a variety of fruit. After breeding, the male molts into olive plumage similar to that of the female. Family groups apparently break up at this time and individuals become solitary and quiet, often sitting motionless in dense foliage. This behavior continues when the birds migrate south to the tropics and is the reason that ornithologists have so little knowledge of exactly where most individuals spend the northern winter.

FACT FILE

RANGE
E North America; winters mainly in N South America E of the Andes

HABITAT
Forests

SIZE
6 in (16 cm)

NORTHERN ORIOLE

Icterus galbula

♂
race *galbula*
"Baltimore Oriole"

Male Northern Orioles are conspicuous, bright orange and black birds that show distinct plumage variations among races. There are 3 distinct forms: Bullock's Oriole *I. g. bullocki* from western North America to Nicaragua, the Baltimore Oriole *I. g. galbula*, shown above, from eastern North America, and the Black-backed Oriole *I. g. abeillei* from south-central Mexico. These forms have recently been treated as separate species. These birds weave nests of pliant plant material with top openings, where the female incubates 4–5 pale blue-gray eggs. The adults feed on insects, fruit, nectar, and spring buds.

FACT FILE

RANGE
Breeds North America; winters in C and N South America

HABITAT
Open and riverside woodland, forest edge, trees near human habitation

SIZE
6–6½ in (16–17 cm)

CHAFFINCH
Fringilla coelebs

♂

race *coelebs*

FACT FILE

RANGE
Europe to W Siberia,
Middle East, N Africa,
Atlantic islands

HABITAT
Woodland, gardens, and
open country with
scattered trees

SIZE
6 in (15 cm)

Over most of its range, the Chaffinch is a very common and familiar bird, often living close to human dwellings. Its simple, far-carrying song varies subtly from one area to another, forming local dialects. There are many races. Those in the north, including *F. c. coelebs*, from much of the European range, tend to have brightly colored males. Races from the drier southern region, such as *F. c. spodiogenys* of Tunisia, are paler and the sexes are more alike. There are also several races on the Atlantic islands, including 3 in the Canaries, which are all heavy-billed and have slate blue on the upperparts. The latter include *F. c. palmae* of Las Palmas Island. Chaffinches form large flocks in winter, feeding on seeds in farmland and coming into gardens to forage. In the breeding season, however, they become highly territorial. The nest is placed in a tree or a bush and is finished on the outside with moss and lichen, often making it difficult to spot.

Western Greenfinch
Carduelis chloris

♂

This stocky finch is well-known in gardens in winter, where it is especially fond of sunflower seeds and peanuts. It will readily take food from hanging dispensers. The more northerly populations migrate in winter, and, at this time, the birds turn up in many different habitats, including sea coasts. During the breeding season, Western Greenfinches take up residence in parks, gardens, and along hedgerows, as well as in larger areas of open woodlands. The characteristic strident, nasal calls readily draw attention to the birds, as does the butterfly-like courtship display flight, which is accompanied by a pleasant song, performed by the male. They build their nests in small trees and bushes, about 7–8 ft (2–2.5 m) above the ground.

FACT FILE

RANGE
Europe, N Africa, Asia Minor, Middle East, C Asia

HABITAT
Open woodland, bushes, gardens

SIZE
6 in (14.5 cm)

AMERICAN GOLDFINCH
Carduelis tristis

These sparkling little finches are commonly seen in flocks feeding on the seeds of roadside thistles and other weeds. When disturbed, the flocks fly up and swirl around, twittering all the time, before eventually settling again. Their sweet, canary-like song, usually given during a circling song-flight, has earned these birds the popular, if ornithologically inaccurate, name of Wild Canary. Though the male American Goldfinch's plumage appears bright and distinctive during the breeding season, it becomes duller in color, and much more like that of the female, through the rest of the year. This species' nest is a neat cup in a tree or shrub.

FACT FILE

RANGE
S Canada, U.S.A.;
N populations migrate
in winter

HABITAT
Open woodland, weedy
fields, roadsides

SIZE
4 in (11 cm)

EURASIAN GOLDFINCH
Carduelis carduelis

♂
race *britannica*

Throughout much of its range, the Eurasian Goldfinch has been heavily trapped as a cage bird because of its bright, delightful plumage and lovely, liquid, twittering song. Trapping has even been blamed for the extinction of some local populations. This species has many races, divided into 2 distinct groups. Western races, such as *C. c britannica* of the British Isles, have black on the head, while eastern races, such as *C. c. paropanisi* of central Asia, have gray. The 2 forms interbreed where they come into contact in central Asia. Like its cousin the American Goldfinch, the Eurasian Goldfinch feeds readily on the seeds of weeds. Flocks of the birds are often seen drifting along roadsides from one patch of thistles to another. The nests are built in bushes or trees, often high up and well out on side branches.

FACT FILE

RANGE
N Africa, Europe, Middle East to C Asia

HABITAT
Open broad-leaved woodland, scrub with weedy areas, gardens

SIZE
5 in (12 cm)

LESSER REDPOLL
Carduelis cabaret

♂
race *cabaret*

Redpolls of several similar forms have recently been considered separate species. The Lesser Redpoll has increased markedly in western Europe, taking advantage of the young conifer growth established by forestry interests. Just a few decades ago, there were only isolated populations of the race in upland Britain and the Alps. Its distribution now runs continuously from Britain, northeast France, and the Low Countries to the Alps and Czechoslovakia, and up into southern Scandinavia. Elsewhere, the Common Redpoll *C. flammea* breeds across northern Eurasia and North America, wintering far south of the breeding range, and the pale Arctic Redpoll *C. hornemanni* breeds near the Arctic tree line. Females lack the pinkish breast and rump of the males; juveniles lack the black chin and red forehead of the adults, or "poll," which gives the species its name.

FACT FILE

RANGE
Circumpolar in N latitudes,
Britain, central Europe

HABITAT
Birch forest, shrub tundra,
young conifer plantations

SIZE
5½ in (14 cm)

House Finch
Carpodacus mexicanus

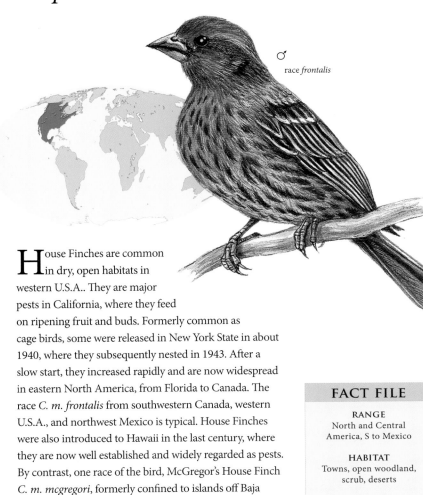

♂
race *frontalis*

House Finches are common in dry, open habitats in western U.S.A.. They are major pests in California, where they feed on ripening fruit and buds. Formerly common as cage birds, some were released in New York State in about 1940, where they subsequently nested in 1943. After a slow start, they increased rapidly and are now widespread in eastern North America, from Florida to Canada. The race *C. m. frontalis* from southwestern Canada, western U.S.A., and northwest Mexico is typical. House Finches were also introduced to Hawaii in the last century, where they are now well established and widely regarded as pests. By contrast, one race of the bird, McGregor's House Finch *C. m. mcgregori*, formerly confined to islands off Baja California, had become very rare by the turn of the century and dwindled to extinction by 1938.

FACT FILE

RANGE
North and Central
America, S to Mexico

HABITAT
Towns, open woodland,
scrub, deserts

SIZE
6 in (15 cm)

EVENING GROSBEAK
Coccothraustes vespertinus

♀

♂

This large, chunky finch was originally (and incorrectly) thought to sing only late in the day, hence its common name. Formerly a bird primarily of the western mountains of North America, the Evening Grosbeak has spread east across Canada as far as Nova Scotia during this century. This expansion of the bird's range is still continuing eastward. During winter, the Evening Grosbeak migrates south and east, often coming to feeding stations, particularly if sunflower seeds are on offer. The numbers occurring each winter vary enormously, depending on the abundance of natural food. The bird's nest is a loose, shallow cup of twigs and roots placed in a woodland tree, sometimes up to 65 ft (20 m) above the ground.

FACT FILE

RANGE
W North America, S to Mexico, E across Canada

HABITAT
Coniferous and mixed woodland

SIZE
8 in (20 cm)

House Sparrow

Passer domesticus

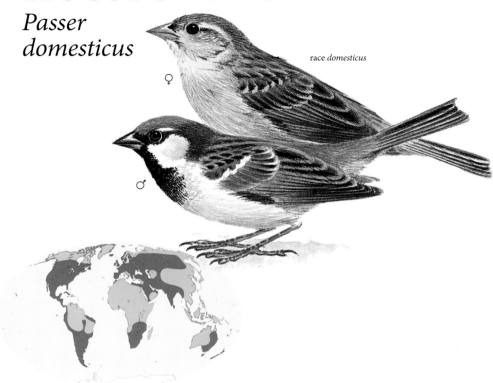

race *domesticus*

♀

♂

The familiar House Sparrow is a seed-eater, adapted to the large seeds of grasses and cultivated grains but also feeding on invertebrates and taking a variety of scraps from bread to meat fibers. The race illustrated is *P. d. domesticus* from much of Eurasia; other races differ in plumage details and size. The House Sparrow is an extremely successful species, nesting in holes or openly in trees in loose colonies and is capable of laying as many as 7 clutches of eggs in a single year in the tropics. In places, the House Sparrow becomes an agricultural pest.

FACT FILE

RANGE
Throughout Eurasia
excluding Far East;
introduced worldwide

HABITAT
Farmland and
built-up areas

SIZE
5½–7 in (14–18 cm)

EUROPEAN STARLING
Sturnus vulgaris

juv

♂

FACT FILE

RANGE
Europe and W Asia;
introduced to North
America, South Africa,
S Australia, New Zealand

HABITAT
Open woods, parks,
gardens, towns

SIZE
8 in (21 cm)

From a distance, the adult has a dark, nondescript appearance, but a closer look reveals a beautiful iridescent adult plumage of blue, violet, green, and bronze. The feathers have pale buff tips, giving the bird a spotted appearance, especially after the annual molt in the fall. By the time the breeding season approaches, these pale tips have become worn away and the plumage is much glossier and, especially in the male, iridescent. Also, the legs change color from brown to reddish-pink and the bill changes from dark brown to yellow, with a creamy pink base in females and a steel-blue base in males. The European Starling is an adaptable opportunist with a broad diet and a complex, variable breeding system, which encompasses monogamy, polygamy, mate exchange, and nest parasitism. This may account for its success as a colonist. In North America, for example, it is one of the most numerous birds, but these are all descendants of some 120 birds released in New York a century ago, which have since increased a millionfold.

BLUE JAY
Cyanocitta cristata

Large noisy groups of food-gathering Blue Jays are a familiar sight in the gardens of North America. They hoard seeds and nuts for the winter, burying them in holes. Despite these precautions, Blue Jays in the far north normally migrate south in winter, whereas the southern birds usually stay near their nesting sites. Although seeds and nuts make up most of its diet, the Blue Jay feeds avidly on insects—particularly the irregular outbreaks of tent moth caterpillars. It will also rob nests of eggs and chicks and take small mammals.

FACT FILE

RANGE
E North America, from
S Canada to Mexican Gulf

HABITAT
Deciduous forest,
semi-woodland

SIZE
12 in (30 cm)

Cardinal Woodpecker *Dendropicos fuscescens*
This woodpecker will probe tree surfaces to
reveal insects and spiders.

Tree Swallow
Tachycineta bicolor
This swallow feeds
mainly on flying insects.

Pied Currawong
Strepera graculina
Berries are favorite foods
for the Pied Currawong.

Sharp-shinned Hawk
Accipiter striatus
This hawk catches and
feeds on small birds.

OPPORTUNIST FEEDERS

Wood Thrush
Hylocichla mustelina
This thrush searches
for food on the
ground.

The birds that visit a backyard depends on the nature of the garden and also on the surrounding area. Close to a wood, woodland predators, such as hawks and owls, may visit, as well as woodpeckers investigating the trees and lawns and foraging flocks of warblers and titmice. If the habitat is right, they will come even if food is not specifically offered as they treat the garden as an extension of the woodland.

A pond, with small fish or frogs, attracts other birds though they may be early-morning raiders that are gone soon after the sun is up, such as herons and kingfishers after an easy meal. Other birds simply come to drink and bathe. Similarly, fruit trees attract birds that eat the fleshy buds or flowers in spring, or the ripening fruit. Waxwings, which are voracious berry eaters in winter, often visit gardens.

Most gardens attract insects, which bring in birds from flycatchers and bluebirds to swallows. In North America, the nighthawk is a possibility and in Africa several related nightjars might fly over. In Europe, it is less likely that a nightjar will visit a garden. A band of magpies is much more likely.

Gray Heron
Ardea cinerea
This heron uses
its long beak
to catch its
prey.

GRAY HERON
Ardea cinerea

Gray herons like waterside habitats, where they catch fish, frogs, and small mammals. They breed farther north than any other heron and some populations suffer high rates of mortality in severe weather. However, the birds have a marked capacity for recovering their numbers. Colony size of the Gray Herons is variable, with about 200 being the maximum but they are usually much smaller. A pair readily lays another clutch of eggs if their first is destroyed—they can repeat this 2–3 times.

FACT FILE

RANGE
Widespread in Eurasia and Africa

HABITAT
Shallow freshwaters of all types; also coasts, especially in winter

SIZE
35½–38½ in (90–98 cm)

MALLARD
Anas platyrhynchos

♀

♂

The success of the Mallard, ancestor of most domestic ducks, reflects its supreme adaptability. It can become completely tame in urban areas, relying on human handouts for food though it is as wild as any wildfowl in other habitats. Mallards feed by dabbling in shallow water or upending to reach greater depths. They are omnivorous, eating both invertebrates and plant matter. Natural nest sites are in thick vegetation close to water, but in towns, the birds use holes in trees and buildings and even window ledges.

FACT FILE

RANGE
N hemisphere, N of the tropics; introduced to Australia and New Zealand

HABITAT
Wide range of fresh and coastal waters

SIZE
20–25½ in (50–65 cm)

EURASIAN KESTREL
Falco tinnunculus

♀

♂

FACT FILE

RANGE
Breeds most of
Europe, Africa, Asia;
N and E birds winter from
Britain S to S Africa and N
India S to Sri Lanka

HABITAT
Diverse, from moors to
tropical savanna and urban
areas; avoids tundra, dense
forest, deserts

SIZE
12–14 in (31–35 cm)

The Eurasian Kestrel is one of the world's
most abundant birds of prey, with a very
wide distribution and the ability to adapt
well to a range of artificial habitats. Although
it is a competent bird-catcher, and feeds widely
on insects and other invertebrates, the Eurasian Kestrel
is best-known as a hunter of small mammals. It frequently
hovers for prolonged periods on fast-beating wings as it
scans the ground. When prey is spotted, the bird descends
in a series of swoops and hoverings, before finally dropping
to snatch the prey in its feet. It also hunts from a high
perch and will sometimes snatch birds in midair. Most
of the 10 or 11 races are similar to the race illustrated,
F. t. tinnunculus from Europe, northwest Africa, and Asia
South to Arabia, Tibet, and extreme southeastern U.S.S.R.

AMERICAN KESTREL
Falco sparverius

race *sparverius*

The American Kestrel feeds on a variety of foods including insects and other invertebrates and small snakes, lizards, birds, and small mammals, including bats. It catches its prey either by swooping down on the creature from a perch or from the air after hovering overhead. One of the brightest-plumaged of all the kestrels, the American Kestrel is unusual in that, even as juveniles, the sexes have distinct plumages. There is considerable geographical variation in size and in plumage in these birds, particularly in that of the male American Kestrel. The race illustrated is *F. s. sparverius*, which can be found in much of North America.

FACT FILE

RANGE
Breeds SE Alaska and Canada S to extreme S South America; N birds winter S to Panama

HABITAT
Open country: mountain meadows, grassland, deserts, open woodland, agricultural land and urban areas

SIZE
9–12 in (23–31 cm)

NORTHERN SPARROWHAWK
Accipiter nisus

♀

♂

FACT FILE

RANGE
Europe and NW Africa E to
Bering Sea and Himalayas

HABITAT
Chiefly woodland or forest

SIZE
11–15 in (28–38 cm)

This widespread hawk eats small birds. Small mammals account for less than 3 percent of the Northern Sparrowhawk's diet. Sparrowhawks are secretive woodland birds but they can hunt in the open, taking prey by surprise in sudden, swift attacks, using their great agility. Male Northern Sparrowhawk's take small prey, such as tits and finches. The larger females often eat thrushes and doves. The female incubates 4–5 eggs in a nest of sticks built

SHARP-SHINNED HAWK
Accipiter striatus

juv

This species shows considerable geographical variation across its range. In North America, males are slate-gray above and closely barred rufous below, while the large females are brown above and barred down below. The race *A. s. velox* found over much of the species' North American range is typical of this type. In stark contrast, the race *A. s. chionogaster* found in Guatemala, Honduras, El Salvador, and Nicaragua is very dark above and white below. There is also some size variation. Races from the West Indies, for instance, are all quite small. In its general behavior, the Sharp-shinned Hawk is similar to most small members of the genus *Accipter*. It is principally a bird-catcher and surprises its prey with a sudden dash from cover.

FACT FILE

RANGE
North, Central, and South America, Caribbean

HABITAT
Woods, forests and other well-timbered areas

SIZE
10–13 in (25–33 cm)

COMMON PHEASANT
Phasianus colchicus

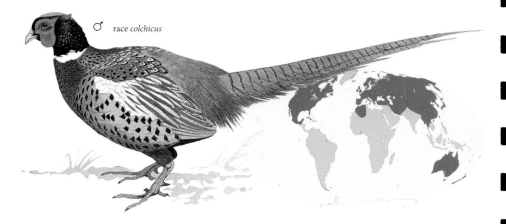

♂ race *colchicus*

FACT FILE

RANGE
Asia; introduced to Europe, Australia, and North America

HABITAT
Lowland farmland, woodland, upland scrub

SIZE
Male 29½–35½ in (75–90 cm); Female 20½–25 in (52–64 cm)

Introduced to many areas for shooting, the Common Pheasant is the most widely distributed of all the gamebirds. The females are well-camouflaged but the colorful males are a frequent and conspicuous sight on open farmland. In Europe, there are now a variety of races, including *P. c. torquatus*, from eastern China, and *P. c. colchicus*, from Transcaucasia and the eastern and southeastern Black Sea region, plus a variety of typical and dark (melanistic) forms. The melanistic form and some others result partly from interbreeding with the Green Pheasant *Phasianus versicolor* from Japan. In its native range (chiefly in India, Southeast Asia, and China), the bird lives on open forest-grassland. It is an opportunist feeder, taking a variety of seeds, grain, fruit, insects, and other small animals, and this is one reason why it has adapted so well to other habitats. The male is polygamous and gathers a harem of females on his display territory.

Common Moorhen
Gallinula chloropus

race *chloropus*

mutual retreat display

Possibly the most abundant species of rail in the world, the Common Moorhen is an unlikely candidate for success, for it is a poor flier and barely adequate swimmer, preferring to search for food on foot with a delicate, high-stepping gait. Males are highly territorial and will fight with their large feet—many are badly injured in the process. The chicks become independent very quickly and often help their parents feed a second brood.

FACT FILE

RANGE
Throughout temperate and tropical Eurasia, Africa, North and South America; but not Australasia

HABITAT
Small ponds, rivers, wet marshes

SIZE
14 in (35 cm)

EASTERN SCREECH OWL
Otus asio

gray phase

rufous phase

FACT FILE

RANGE
E North America from extreme S Quebec and Ontario to E Montana and C Texas and from S Canada to the tip of Florida and NE Mexico

HABITAT
Open forest, riverside woodland, orchards and other cultivated land, parks, and gardens with large trees

SIZE
7½–9 in (19–23 cm)

There are 2 color phases of this owl—a rust-red rufous phase and a gray phase. The ear tufts are conspicuous when erect, but may be held against the head where they are nearly invisible. Like many nocturnal owls, they are more often heard than seen. Their primary song is a tremulous descending whinny. They also utter a quickly repeated, monotonous *who-who-who* all on one pitch. Twitters and barks are given in agitation and in response to predators. A very close relative, the Western Screech Owl *O. kennicottii*, found to the west of the Rockies, is similar in appearance but can be distinguished by its different songs—a series of short whistles accelerating in tempo and a brief trill followed by a longer one. This species feeds on large invertebrates, fish, amphibians, reptiles, birds, and mammals. Small mammals and birds form their chief prey but they sometimes kill birds larger than themselves.

GREAT HORNED OWL
Bubo virginianus

race *virginianus*

race *pallescens*

The Great Horned Owl is large and relatively conspicuous. It varies in size and color according to habitat. Those Great Horned Owls from humid forests are dark while desert birds are paler. The race *B. v. virginianus* from eastern North America is a dark form; *B. v. pallescens* is a pale race from the deserts of Arizona and Mexico. Great Horned Owls make various booming hoots and barks. They take over old stick nests or use cavities in cliffs or buildings for their 2–3 eggs, which are incubated only by the female. The young leave the nest after 5–6 weeks. Great Horned Owls feed on mammals and birds up to the size of jack rabbits, as well as almost anything else they can catch, from insects to fish and lizards.

FACT FILE

RANGE
North America, except tundra regions of Alaska and Canada, S to Tierra del Fuego

HABITAT
Lowlands to tree line in deciduous or coniferous boreal, temperate, and tropical forests, prairies, deserts, farmland, and occasionally suburban areas

SIZE
17–21 in (43–53 cm)

Northern Saw-whet Owl

Aegolius acadicus

juv

This relatively tame little owl is a smaller version of the more widespread Tengmalm's, or Boreal, Owl of northern Europe, Siberia, Alaska, and Canada. The 2 species overlap in a 125 ml (200 km) wide zone in southern Canada and the northern Midwest U.S.A., and it is still not completely understood how they can coexist. The odd name refers to the bird's song, which resembles the sound of a saw blade being sharpened. Northern Saw-whet Owls also utter a variety of whistles, single notes, and rasping sounds. They eat mainly mice and voles, supplemented with the occasional small bird, and lay their eggs in abandoned woodpecker holes.

BOOBOOK OWL
Ninox novaeseelandiae

race *ocellata*

race *leucopsis*

This common small brown owl's familiar double hoot, from which it takes its name, is reminiscent of the call of the European Cuckoo *Cuculus canorus*. Some of the 14 or so races have been regarded as separate species. There are 4 Australian races, including the pale rufous or sandy-brown *N. n. ocellata* from northern, inland, and much of western Australia and the darker, heavily spotted *N. n. leucopsis* from Tasmania. Boobook Owls feed mainly on insects, sometimes caught around town lights, and also on small mammals, birds, and geckos. By day they roost in thick foliage or sometimes in tree hollows, caves, or buildings.

FACT FILE

RANGE
Australia, Tasmania, Lesser
Sundas, Moluccas, S New
Guinea, Norfolk Island,
New Zealand

HABITAT
Forest, woodland,
scrub, urban trees

SIZE
10–14 in (25–35 cm)

COMMON GROUND-DOVE
Columbina passerina

ommon Ground-doves are tiny birds, little larger than
sparrows. They usually occur in pairs, though they will
often gather in flocks in winter. Largely terrestrial, they forage
on the ground, bobbing their heads as they walk. Small seeds
and berries make up the bulk of their diet, along with the
occasional insect. Their usual call is a soft, cooing whistle.
When they do fly, their bright chestnut primary wing feathers
are apparent. The female is generally more dull than the male.
Courtship displays are elaborate, with the male strutting about
before the female or chasing her, cooing as he does so and
puffing his feathers out. The birds nest mostly on the ground,
but will also use brushes, palm fronds, vines, and low shrubs.
The flimsy cup is made from grasses and plant fibers. The pair
sometimes reuse the nest for a second brood and occasionally
use the old nest of another species as a platform for their own.

FACT FILE

RANGE
S U.S.A. and Caribbean to
Ecuador and Brazil; Bahamas

HABITAT
Open country, roadsides,
farms, grassland

SIZE
6½ in (16.5 cm)

INCA DOVE
Scardafella inca

This tiny dove reveals the rich chestnut color of its primary wing feathers when it takes flight. Its usual call is a series of 2-note coos, but it also utters a variety of soft or throaty display, threat, and alarm calls. Inca Doves form flocks in winter and roost communally. On cold nights, they are known to huddle together in groups, with some birds on top of the others forming "pyramids." They are largely terrestrial feeders, foraging mainly for seeds and cultivated grain. As with all pigeons, the young are fed on crop "milk"—a liquid food secreted in the esophagus of the adults. Inca Doves are monogamous, with an elaborate courtship display in which the male struts, head-bobs, fans his tail, and utters cooing notes. They nest in bushes or shady trees, 3–23 ft (1–7 m) above the ground.

FACT FILE

RANGE
Locally from SW U.S.A.
to N Costa Rica

HABITAT
Towns, open areas,
farms, riverside scrub,
cactus, mesquite

SIZE
8 in (20 cm)

SULPHUR-CRESTED COCKATOO
Cacatua galerita

♂

Common throughout most of its range, this handsome, noisy parrot often associates in large flocks outside of the breeding season, descending on grasses, shrubs, and trees to feed on seeds, fruit, palm hearts, and insects. It can even be seen in urban areas. It is an occasional pest in crops and is sometimes shot for this reason. Sulphur-crested Cockatoos also suffer as a result of hunting in New Guinea. While the flock is feeding on the ground, a few birds stand sentinel in nearby trees and warn of danger with loud raucous cries. At dusk, the flock returns to a habitual roost. The flocks break up during the breeding season. Breeding pairs nest in vertical cavities in eucalypts, usually near water, and both parents incubate the 2–3 eggs.

FACT FILE

RANGE
Melanesia, New Guinea, N and E Australia

HABITAT
Lowland forest, savanna and partly cleared land up to 5,000 ft (1,500 m)

SIZE
15–19½ in (38–50 cm)

YELLOW-BILLED CUCKOO

Coccyzus americanus

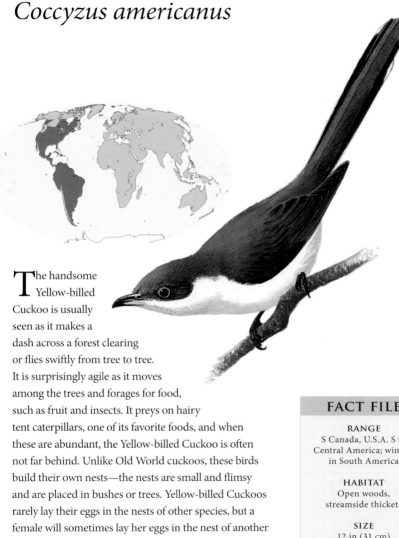

The handsome Yellow-billed Cuckoo is usually seen as it makes a dash across a forest clearing or flies swiftly from tree to tree. It is surprisingly agile as it moves among the trees and forages for food, such as fruit and insects. It preys on hairy tent caterpillars, one of its favorite foods, and when these are abundant, the Yellow-billed Cuckoo is often not far behind. Unlike Old World cuckoos, these birds build their own nests—the nests are small and flimsy and are placed in bushes or trees. Yellow-billed Cuckoos rarely lay their eggs in the nests of other species, but a female will sometimes lay her eggs in the nest of another Yellow-billed Cuckoo.

FACT FILE

RANGE
S Canada, U.S.A. S to Central America; winters in South America

HABITAT
Open woods, streamside thickets

SIZE
12 in (31 cm)

GALAH
Eolophus roseicapillus

♂

This is the most common cockatoo—indeed, one of the most common of all parrots—in Australia. It is one of the few species to have benefited from the European settlement, having increased its range and numbers in response to forest clearance, crop planting, and the provision of water points. It is considered a pest by cereal growers, for it takes grain as well as grass seeds, fruit, green shoots, and insects. The Galah is sometimes known as the Roseate Cockatoo. The nest is usually in a vertical tree hollow, though holes in cliffs are sometimes used. The pairs bond for life and defend the nest-hollow against intruders. Both male and female share the incubation and feeding of the 2–6 young. The newly fledged birds gather into treetop crèches of up to 100 birds, which eagerly await the return of their parents with food. After 6–8 weeks, the young are left to fend for themselves while their parents go away to molt. Young Galahs spend their first 2 or 3 years among large wandering flocks of non-breeding birds.

FACT FILE

RANGE
Most of Australia

HABITAT
Eucalypt woodland,
watercourse vegetation,
and grassland

SIZE
14 in (35 cm)

COMMON NIGHTHAWK
Chordeiles minor

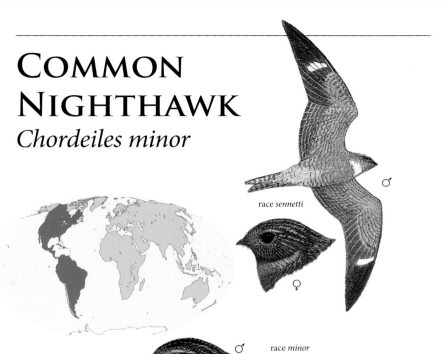

race *sennetti*

♂

♀

♂   race *minor*

The white patches on the sharply pointed wings of the Common Nighthawk are often its most conspicuous feature as it swoops and wheels through the twilight in pursuit of flying insects. It will eat moths, flies, grasshoppers, beetles, mosquitoes, and ants; indeed, one bird was found to have the remains of 2,175 ants in its stomach. The male performs a dramatic courtship ritual, swooping down with booming wings on to a perch near a potential mate, then displaying on the ground. The bird's color varies from dark brown in the eastern North American race *C. m. minor* to paler grayish brown in the Great Plains race *C. m. sennetti*—these color differences are subtle in adults but striking in juveniles.

FACT FILE

RANGE
Breeds in North America (except extreme N) and West Indies; winters in South America S to Argentina

HABITAT
Open fields, gravelly areas, grassland, savanna, semi-desert, towns

SIZE
9½ in (24 cm)

RIVER KINGFISHER
Alcedo atthis

race *ispida*

♀

♂

race *hispidoides*
in nest-burrow

FACT FILE

RANGE
Breeds Europe, NW Africa,
Asia, Indonesia to Solomon
Islands; winters in S of range

HABITAT
Clear, slow-moving streams
and small rivers, canals,
ditches, reeds, marshes;
coasts in winter

SIZE
6 in (16 cm)

This bird and the Belted Kingfisher are the most northerly breeding species of kingfishers. Along with the European race *A. a. ispida*, the illustration shows *A. a. hispidoides*, which ranges from northern Sulawesi to the Bismarck Archipelago off Papua New Guinea. The River Kingfisher is aggressively territorial, defending a ½–3 ml (1–5 km) stretch of stream in winter, even from its mate. In spring, it will drive away both rivals and small songbirds. It excavates its burrow in the bank of stream, river, or gravel pit, usually above water; most nest-burrows are 18–36 in (45–90 cm) long. It rarely feeds away from water and is an expert fisher.

BELTED KINGFISHER
Megaceryle alcyon

This solitary kingfisher prefers clear waters with overhanging trees, wires, or other perches. It usually hovers and dives for fish, but will also prey on insects, crayfish, frogs, snakes, and even the occasional mouse. A loud, dry, rattling call often announces the presence of this bird. The breeding distribution of the Belted Kingfisher seems to be limited only by the availability of foraging sites and earth banks in which it can excavate its nest chamber. This lies at the end of an upward-sloping tunnel 3–16 ft (1–5 m) long.

FACT FILE

RANGE
Breeds through most of North America; winters in ice-free areas S to West Indies and N Colombia

HABITAT
Streams, rivers, ponds, marshes

SIZE
12 in (30 cm)

HOOPOE
Upupa epops

race *epops*

FACT FILE

RANGE
Europe, Asia, Africa,
Madagascar; N birds winter
in the tropics

HABITAT
Woodland, savanna, parks,
lawns, orchards, farmland

SIZE
11 in (28 cm)

The Hoopoe's striking plumage, its mobile crest, its far-carrying *hoo-poo-poo* call, and its association with human settlement make it a familiar bird. Daytime migrants are so conspicuous that, despite pungent, defensive secretions, they are a favorite prey of Eleanora's and Sooty Falcons. The nest is a thinly lined cavity in a tree or wall or in the ground. The eggs vary in color from gray to pale yellow or olive and become heavily stained in the nest. The clutch size varies from 2–9 eggs, and incubation takes 15–16 days. The nestling period lasts about 4 weeks.

SPECKLED MOUSEBIRD
Colius striatus

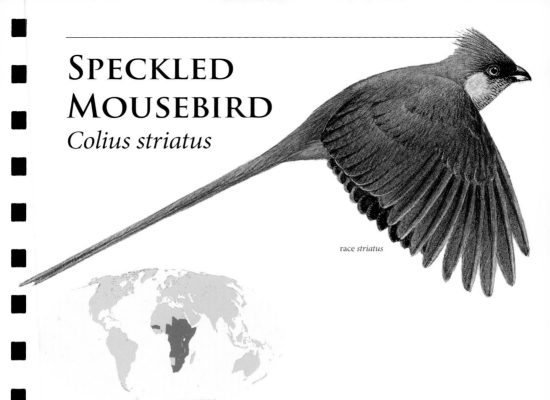

race *striatus*

Mousebirds take their name from their striking, rodentlike appearance as they creep and scramble through vegetation. Members of family groups follow each other with alternate whirring and gliding flight, their extremely long tails, up to 8½ in (22 cm) long, projecting stiffly behind them. They often perch in clusters and more than 20 birds may roost together in a tight bunch. There are some 17 races: *C. s. striatus* from southwestern South Africa is illustrated. Speckled Mousebirds use their robust, arched, hooked bills to feed on all kinds of vegetable matter. Flocks of them can cause much damage in gardens and orchards.

FACT FILE

RANGE
C, E, and S Africa

HABITAT
Woodland, scrub, gardens, hedges

SIZE
12–14 in (30–36 cm)

LAUGHING KOOKABURRA
Dacelo novaeguineae

race *novaeguineae*

♂

FACT FILE

RANGE
E and SE Australia;
introduced to Tasmania
and SW Australia

HABITAT
Open forest, urban trees

SIZE
16–18 in (40–45 cm)

This giant kingfisher has a loud, rollicking territorial call that sounds like maniacal human laughter. It also utters softer chuckles. There are just 2 races: *D. n. novaeguineae*, is the widespread and well-known bird of the eastern half of Australia, while the other, smaller race *D. n. minor* is confined to the Cape York Peninsula. Laughing Kookaburras feed by pouncing on a variety of terrestrial invertebrates, reptiles, small mammals, birds, and nestlings. They can even devour snakes up to 3 ft (1 m) in length. Although they are not closely associated with water, they will sometimes catch fish with plunging dives and, on occasion, raid suburban goldfish ponds. They live in family groups that defend the same territories all year round.

BLUE-THROATED BARBET
Megalaima asiatica

race *davisoni*

The Blue-throated Barbet is common from the Himalayan foothills down to low level plains. It prefers groves of fig trees around villages and wooded slopes, where it searches for fruit. It also eats large insects, which it strikes against a branch before swallowing. The race illustrated is *M. a. davisoni*, which ranges from southern Burma to southern China and northern Indochina. It is a noisy species, which often calls together with other barbets, making a confusion of sounds. Pairs may call in such rapid succession that their duets sound as if they are coming from a single bird. The nest-hole is excavated in a tree branch and the cavity may be used for several years, although a fresh entrance is often cut each season.

FACT FILE

RANGE
N India, Southeast Asia

HABITAT
Light mixed forest, gardens

SIZE
9 in (23 cm)

CARDINAL WOODPECKER
Dendropicos fuscescens

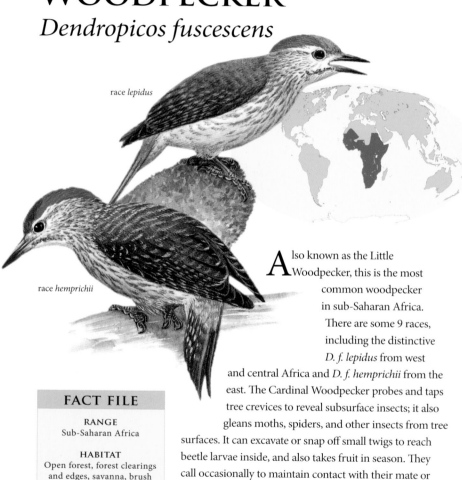

race *lepidus*

race *hemprichii*

FACT FILE

RANGE
Sub-Saharan Africa

HABITAT
Open forest, forest clearings
and edges, savanna, brush

SIZE
5–6 in (13–15 cm)

Also known as the Little Woodpecker, this is the most common woodpecker in sub-Saharan Africa. There are some 9 races, including the distinctive *D. f. lepidus* from west and central Africa and *D. f. hemprichii* from the east. The Cardinal Woodpecker probes and taps tree crevices to reveal subsurface insects; it also gleans moths, spiders, and other insects from tree surfaces. It can excavate or snap off small twigs to reach beetle larvae inside, and also takes fruit in season. They call occasionally to maintain contact with their mate or family members and use a tinny rattling note for territorial defense; both sexes drum softly. They may also join foraging parties of other bird species.

Northern Wryneck
Jynx torquilla

The Northern Wryneck can twist and writhe its neck—hence its common name. These contortions resemble the movements of a snake and, when combined with snakelike hissing sounds, serve as a deterrent to small predators. Northern Wrynecks feed on ants and other insects, on the ground or in trees, but do not use their tails as a brace. Nor do they produce the drumming characteristic of true woodpeckers and piculets. Wrynecks use tree holes for nesting, often evicting other birds from their nests. These birds are classified in a subfamily of their own. They have one close relative, the Rufous-necked Wryneck *J. ruficollis*.

FACT FILE

RANGE
Breeds Eurasia to N Africa;
winters C Africa and S Asia

HABITAT
Open forest, parkland

SIZE
6–6½ in (16–17 cm)

GREEN WOODPECKER
Picus viridis

♂

race *viridis*

FACT FILE

RANGE
W Eurasia from S Scandinavia, Europe to mountains of N Africa, Turkey, Iran, W Russia

HABITAT
Deciduous and mixed forest edged, secondary growth, parklike habitats

SIZE
12 in (30 cm)

This bird spends much of its time foraging on the ground for ants and grubs. When it finds an anthill, the Green Woodpecker tears into it with a vigorous twisting of the head and probing of the beak. It also gleans insects and spiders from tree surfaces, consumes fruit and seeds, and sometimes raids beehives. The female has an all-black "mustache," while juveniles have only a faint one and barred underparts. There are several races, including *P. v. viridis* of Europe and *P. v. vaillantii* of north Africa, which has strongly barred underparts and less red on the head in the female, whose "mustache" resembles that of the male. The Green Woodpecker tends to be solitary outside of the breeding season. Both sexes have similar calls, including a far-carrying laughing cry and a series of *kwit-up* notes with variations. This species rarely drums.

BOHEMIAN WAXWING
Bombycilla garrulus

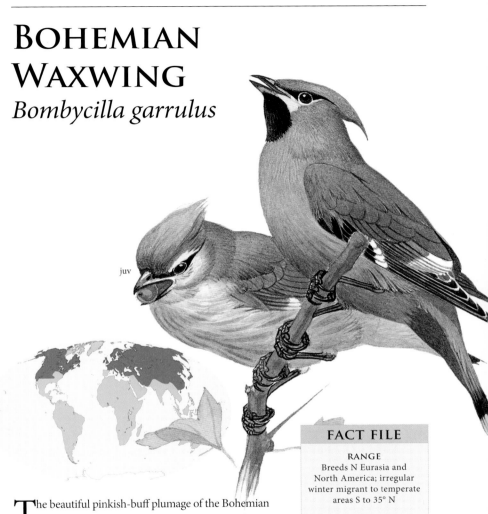

juv

The beautiful pinkish-buff plumage of the Bohemian Waxwing is offset by yellow tips to the tail feathers, and waxlike red blobs at the tips of the secondary wing feathers, from which the common name of this species is derived. During their courtship display, the male presents the female with a berry or ant pupa, which the birds then pass back and forth from beak to beak but never swallow.

FACT FILE

RANGE
Breeds N Eurasia and North America; irregular winter migrant to temperate areas S to 35° N

HABITAT
Breeds in dense coniferous or mixed forest; occasionally found in town parks and gardens in winter

SIZE
8 in (20 cm)

TREE SWALLOW
Tachycineta bicolor

FACT FILE

RANGE
Breeds C Alaska and Canada to CE U.S.A.; winters S U.S.A., Caribbean, Central America

HABITAT
Open areas near water, especially flooded areas with dead trees

SIZE
5–6 in (12.5–15 cm)

During migration, this species forms huge flocks that often roost in the low vegetation of coastal marshes. Though it relies primarily on a diet of flying insects, in extreme cold it has been observed feeding on bayberries. The Tree Swallow nests in ready-made holes, choosing old woodpecker cavities, crevices, and nest-boxes in which to build its cup-shaped nest of dried grasses or pine needles. This is then lined with white feathers. The competition for feathers is keen and adults have been known to pluck feathers from the backs of domestic ducks. The usual clutch contains 4–6 white eggs and extra birds may occasionally assist the parents in caring for the brood.

Common Koel
Eudynamys scolopacea

race *scolopacea*

♂

♀

race *cyanocephala*

♀

In the breeding season, this cuckoo can often be heard calling—the male uttering loud, shrill whistles and the characteristic *ko-el* call, while the female responds with short, piercing whistles. Like most cuckoos, they lay their eggs in the nests of other birds—in India, the hosts are crows, while in Australia, they include large honeyeaters and orioles. The chick takes all the food and either starves out or ejects the other nestlings. The adults forage for fruit in the forest canopy and often visit gardens to steal cultivated fruit. There are 15 or more races of the Common Koel, including the race *E. s. scolopacea* from India, Sri Lanka, and the Nicobar Islands and *E. s. cyanocephala* from eastern Australia, which is sometimes considered to be a separate species.

FACT FILE

RANGE
N and E Australia, India, Southeast Asia to New Guinea and Solomon Islands

HABITAT
Forests with fruiting trees

SIZE
15–18 in (39–46 cm)

GRAY CATBIRD
Dumetella carolinensis

The Gray Catbird often forages on the ground and has a habit of flinging leaves about during its search for food. Its diet includes a variety of fruit and invertebrates. Since it can survive on fruit alone, it can winter occasionally in some northern areas. This bird is well-known for the catlike cries that give it its name and is a versatile mimic. A nocturnal migrant, the Gray Catbird often arrives in the southern U.S.A. in spring in large waves. Nesting activities begin soon after arrival. The female builds a low, well-concealed nest in dense thickets, bushes, small trees, or tangles of vines. The male provides some help by bringing materials, such as twigs, strips of bark, leaves, grasses, and even paper and string. They usually raise 2 broods in one season. The nestlings of the first brood are fed mostly insects, while those of the second brood are given much fruit to eat. Gray Catbirds are aggressive in defense of their nests and young.

FACT FILE

RANGE
Breeds S Canada, U.S.A.;
winters Central America,
West Indies

HABITAT
Dense vegetation in
woodland edge;
readily adapts to
human settlements

SIZE
8 in (20 cm)

Brown Thrasher
Toxostoma rufum

This well-known species is most commonly observed skulking along the ground in search of food, or singing on an exposed perch. It typically forages by poking through leaves and other ground cover with its bill, occasionally pausing to pick up leaves and toss them aside. It feeds mostly on insects, but also eats other invertebrates and some small vertebrates. Fruit and acorns also contribute to its diet, especially in winter. Breeding begins as early as March in the southern part of the range and continues through July. The male sings from the top of a tree or tall shrub, head held high, and long tail drooping. His song is a series of rapid, short, musical phrases, usually uttered in pairs. The song resembles that of the Northern Mockingbird, but is wilder. The nest is a bulky cup of sticks, lined with dead leaves, grasses, and fine rootlets. It is often placed in thorny bushes to provide added protection.

FACT FILE

RANGE
E U.S.A. and S Canada
to the foothills of the
Rocky Mountains

HABITAT
Woodland, forest edge,
hedgerows, scrubland,
pastures, gardens

SIZE
10 in (25 cm)

Wood Thrush
Hylocichla mustelina

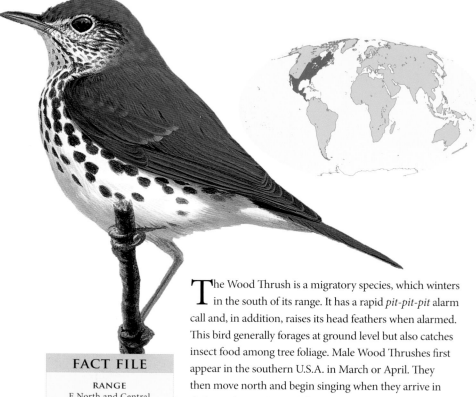

FACT FILE

RANGE
E North and Central America, from S Canada to Panama

HABITAT
Woodland, wooded slopes, parks, often near streams

SIZE
7½–8½ in (19–22 cm)

The Wood Thrush is a migratory species, which winters in the south of its range. It has a rapid *pit-pit-pit* alarm call and, in addition, raises its head feathers when alarmed. This bird generally forages at ground level but also catches insect food among tree foliage. Male Wood Thrushes first appear in the southern U.S.A. in March or April. They then move north and begin singing when they arrive in their nesting territories. The song consists of loud, flutelike phrases with 3–5 notes, each differing in pitch and ending with a soft, guttural trill. The female Wood Thrush builds a bulky cup nest of dead leaves, moss, and rootlets, the lack of grass distinguishes it from the nest of an American Robin. A piece of white paper or cloth is often included in the base of the nest.

FIELDFARE
Turdus pilaris

juv

This is a gregarious, noisy species of thrush that has colonized parts of western Europe, including parts of the British Isles, in recent years. It is essentially a migratory species that breeds in northern latitudes and usually only winters in the southern part of its range. The Fieldfare eats many kinds of invertebrates, as well as plant food, such as seeds and fruit. Fruit is especially important in the fall and winter, and the bird is a specialist feeder on apples and haws. The nest is a bulky cup of grass reinforced with roots and mud, usually placed in the fork of a tree. The typical clutch consists of 5 –6 pale blue eggs with brown markings. When defending their young, the adults will readily and accurately "bomb" predators with their droppings.

FACT FILE

RANGE
Breeds N Eurasia E to C Siberia, Greenland; winters widely in Europe, SW Asia

HABITAT
Subarctic scrub, light coniferous and birch woodland, parks, gardens, towns; winters in open country, woodland edges, fields

SIZE
10 in (26 cm)

EURASIAN JAY
Garrulus glandarius

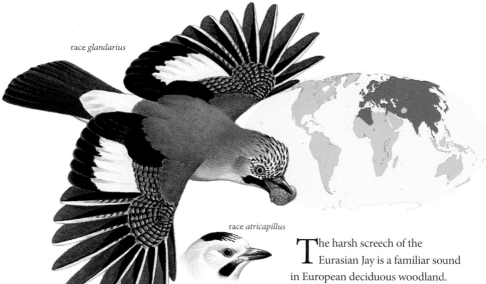

race *glandarius*

race *atricapillus*

race *bispecularis*

FACT FILE

RANGE
W Europe, across Asia to Japan and Southeast Asia

HABITAT
Oakwoods, open country with oaks

SIZE
13 in (33 cm)

The harsh screech of the Eurasian Jay is a familiar sound in European deciduous woodland. The first note is often followed by a second and a third as a family party scours the trees for insects, beech nuts, and acorns. This handsome bird has played a major part in shaping the landscape. Its habit of carrying acorns long distances (especially uphill) and burying them has been an important factor in the dispersal of oak trees and it is likely that the acorn has evolved to attract the Eurasian Jay. For their part, the birds rely heavily on stored acorns for winter food. Of the many races, those illustrated are *G. g. glandarius* of Europe, *G. g. atricapillus* of the Middle East, and *G. g. bispecularis* of the western Himalayas.

COMMON MYNAH
Acridotheres tristis

The Common Mynah has learned how to profit from association with humans and as human habitation spreads, so does the range of the bird. It has a celebrated delectation for locusts and the bird has been introduced to many places, especially tropical islands, in the hope that it would control insect pests. Unfortunately, the bird has become a pest itself in most areas, by eating fruit. This jaunty bird shows a conspicuous white wing patch in flight and often makes its presence felt through its loud fluty calls. Daytime flocks of Common Mynahs assemble in large communal roosts at night, and the large trees in which the birds sleep are the stage for a raucous musical chorus at dawn and dusk.

FACT FILE

RANGE
Afghanistan, India, and Sri Lanka to Burma; currently extending its range to Malaysia and U.S.S.R.; introduced to South Africa, New Zealand, Australia

HABITAT
Farmland, parks, towns

SIZE
9 in (23 cm)

PIED CURRAWONG
Strepera graculina

This species is primarily a tree-dweller. The name currawong comes from one of the bird's calls, but it utters a wide variety of other loud and often musical sounds. Pied Currawongs tend to inhabit forests in the breeding season, where they snatch stick insects and other large arthropods from the foliage. They are also nest-robbers, and may have a major impact on the breeding success of other birds. In winter, large numbers move into towns, where they feed on berries. Their numbers have increased with the introduction of many new berry-producing plants; by dispersing seeds, the birds have helped to spread these exotic plants.

FACT FILE

RANGE
E Australia

HABITAT
Forest and woodland

SIZE
18 in (46 cm)

MAGPIE-LARK
Grallina cyanoleuca

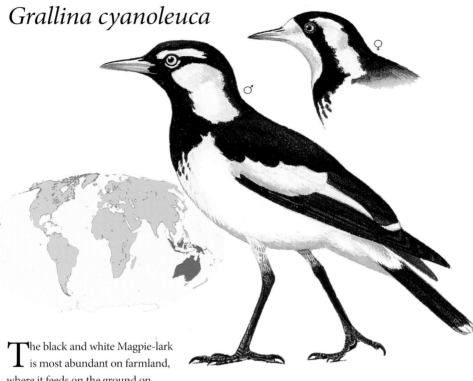

♂

♀

The black and white Magpie-lark is most abundant on farmland, where it feeds on the ground on insects, snails, and other invertebrates. It also frequents the edges of lakes and ponds and has become common in towns, where it is often seen feeding alongside busy roads, seemingly oblivious to the traffic. It has a loud *pee-wee* call, which gives it one of its popular names. The clearance of forests and the provision of water in dry regions have probably encouraged this species. However, its population has declined in some areas, perhaps because the water has become too saline or because the snails have been poisoned in an effort to control liver flukes.

FACT FILE
RANGE
Australia
HABITAT
Open pasture and grassland
SIZE
10½ in (27 cm)

FIGBIRD
Sphecotheres viridis

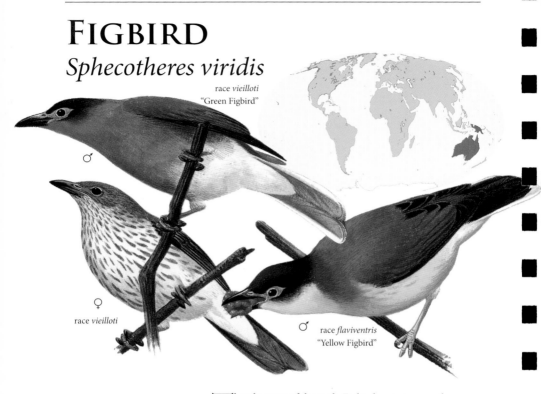

race *vieilloti*
"Green Figbird"

♂

♀

race *vieilloti*

♂ race *flaviventris*
"Yellow Figbird"

FACT FILE

RANGE
N and E Australia,
New Guinea

HABITAT
Rain forest, eucalypt forest,
parks, gardens

SIZE
11 in (28 cm)

The plumage of the male Figbird varies across the bird's range. In the northern race *S. v. flaviventris*, the underparts are yellow, while the Green Figbird *S. v. vieilloti* from eastern Australia has a green belly and a gray throat. Hybridization occurs where these 2 forms meet, and birds with yellow, gray, green, or white throats may be seen. The natural habitat of the Figbird is forest, but it has become one of the most common urban birds in many parts of its range. It eats a wide variety of fruit, both wild and cultivated, as well as some insects. The shallow nest is composed of thin twigs, grass and plant tendrils, placed toward the end of a slender branch, and is often so flimsy that the eggs are visible inside against the light.

BARRED ANTSHRIKE
Thamnophilus doliatus

♀

♂

race *intermedius*

This species is typically found in semi-open areas, such as dry forests or bushy savannas, but it is widespread where it has few competitors. In the heart of Amazonia where antbirds abound, however, it is confined to shrubby growth on riverbanks and islands. The Barred Antshrike searches for insects primarily in the dim light of often impenetrable thickets. More often heard than seen, members of a pair keep in touch through a large repertoire of ends in an emphatic nasal note. As it sings, the Barred Antshrike stretches its neck, raises its striking crest, and vibrates its slightly fanned tail. The amount of black barring on the underparts of the male varies between races.

FACT FILE
RANGE
Mexico to extreme
N Argentina
HABITAT
Lowland thickets and tangles
SIZE
5½–6½ in (14–17 cm)

EASTERN KINGBIRD
Tyrannus tyrannus

FACT FILE

RANGE
Breeds S Canada through U.S.A. to the Gulf coast; winters in Central and South America

HABITAT
Open areas, woodland edge, streamsides, orchards

SIZE
8–9 in (21–23 cm)

Celebrated for its dauntless defense of its breeding territory, the male Eastern Kingbird will attack anything that enters its airspace. Its main targets are larger birds, such as crows and hawks, which are chased relentlessly and rained with blows. At times, this species will even land on its flying victim to knock it about more effectively. Its noisy, blustering calls typify its temperament. Outside the breeding season, it often forages in small, loose flocks. It feeds mainly on flying insects taken on the wing, but will also pluck berries from bushes while hovering. At dusk, it retreats to a communal roost, which may number hundreds or even thousands of birds.

COMMON GRACKLE
Quiscalus quiscula

♂

The Common Grackle forms large, noisy roosts, often containing thousands of birds uttering loud *chuck* calls. The short, harsh, squeaky song sounds like the noise of a rusty hinged gate. These birds nest in small colonies. The females build bulky nests of sticks and grasses 3–13 ft (1–4 m) above the ground. The nest is lined with paper, string, rags, and other debris. Young birds eat insects and spiders; adults eat a wide range of animals, from earthworms and insects to fish, frogs, mice, and the eggs and young of small birds. They also relish seeds and grain and large flocks of grackles can damage crops.

FACT FILE

RANGE
E U.S.A.; N populations
winter S as far as Florida

HABITAT
Open woodland, scattered
trees, forest edge, areas near
human habitation

SIZE
11–13 in (28–33 cm)

RED-EYED VIREO
Vireo olivaceus

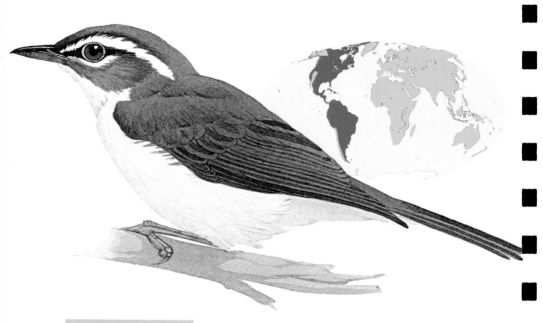

FACT FILE

RANGE
Breeds Canada S to the
Gulf States; winters from
NW South America
to Amazonia

HABITAT
Deciduous or mixed forest,
riverbank woodland

SIZE
5½ in (14 cm)

The Red-eyed Vireo's melodius and fairly rapid song is composed of 2–3 syllables. It is highly vocal and strongly territorial in its summer breeding grounds, where it gleans treetop foliage for insects to eat. By contrast, in its South American winter quarters the Red-eyed Vireo does not sing and eats fruit rather than insects. During the breeding season, the male displays by swaying with body feathers fluffed and tail fanned, to which females may respond with a call.

AMETHYST STARLING
Cinnyricinclus leucogaster

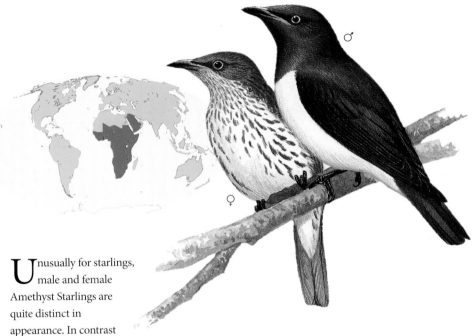

♂

♀

Unusually for starlings, male and female Amethyst Starlings are quite distinct in appearance. In contrast to the gaudy violet and white male, the female, like the juvenile, has mainly brown plumage. They are also less noisy than other species, despite being highly gregarious. They nest in tree holes, lining the cavity with feathers, soft plant material, and even animal dung. In some parts of their range, these starlings are year-round residents. Elsewhere they are nomadic, or even migratory, although they do not travel far. They have a taste for fruit and flocks of these birds will arrive in an area when the fruit ripens— they will then strip the trees and disappear. They will also take insects, often hawking for termites on the wing.

FACT FILE

RANGE
SW Arabia,
sub-Saharan Africa,
South Africa, Gabon,
Congo, and Zaire

HABITAT
Rain forest, wooded
savanna, woodland edge,
parks, gardens

SIZE
6–7 in (16–18 cm)

SUPERB STARLING
Spreo superbus

FACT FILE

RANGE
SE Sudan, E to
Somalia and
S to Tanzania

HABITAT
Open acacia savanna,
lawns, farmland

SIZE
7 in (18 cm)

This spectacular-looking starling is a common East African bird. It feeds mainly on the ground, eating a range of seeds, fruit, and insects. It will devour food scraps with relish and small flocks will often gather at campsites and hotels to beg for food, which makes it very popular with tourists. While Superb Starlings will nest in holes like other starlings, they often build large, untidy, domed nests in low thorn bushes. Occasionally a breeding pair is assisted by one or more non-breeding birds which help to feed the chicks. Juveniles can be distinguished by their duller plumage and brown eyes.

PIED CROW

Corvus albus

Although most of its diet consists of seeds and insects taken from the ground, the Pied Crow is most commonly seen congregating around carcasses or garbage dumps. It is a bold bird with little fear of humans or larger predators and it is an eager mobber of African eagles and buzzards. Its young stay in the nest a long time—they take 40 days to fledge—and the parents often fail to keep up the food supply. As a result, few pairs raise more than 3 young out of a clutch of up to 6 eggs. Many are also parasitized by Great Spotted Cuckoos. The cuckoo chicks hatch sooner and are more vigorous and aggressive than the Pied Crow chicks, which may starve, although usually some are raised alongside the intruders.

FACT FILE

RANGE
Sub-Saharan Africa,
Madagascar

HABITAT
Open and semi-open
country, forest clearings,
urban areas

SIZE
18 in (45 cm)

Barn Swallow *Hirundo rustica*
This swallow will attach its nest to barn walls and other structures.

Common Barn-owl
Tyto alba
Barns and outbuildings make good nesting sites for this owl.

Eastern Bluebird
Sialia sialis
Next-boxes have been important for this bird.

Pied Wagtail
Motacilla alba
This wagtail will roost in heated greenhouses.

GARDEN AND HOUSE NESTERS

Northern Mockingbird
Mimus polyglottos
This bird will often nest near human dwellings.

Our homes provide nesting places for many birds, from gulls on flat roofs to swifts, starlings, sparrows, and martins in the roof or under the eaves. Large, older buildings in more rural surroundings may have owls in the roof space, though they may prefer outhouses to occupied houses. The chimney swift lives up to its name in North America, while in Europe, large chimneys are more likely to be blocked by nesting jackdaws.

In the garden, backyard nesters take advantage of the safety of thick shrubs and hedgerows—thrushes, such as the blackbird in Europe and the American robin are typical. In the U.S.A., eastern bluebirds have been helped to recover their numbers by nest-box schemes. Multi-story towers have big colonies of purple martins. Single boxes are ideal for sparrows, blue and great tits, and starlings.

In gardens with taller trees, there may be natural holes, or bigger boxes can be put up for birds as varied as kestrels, barn owls, and doves. These boxes can succeed, if the birds can find enough natural food for their family.

Winter (Northern) Wren
Troglodytes troglodytes
This wren is at home in gardens.

HERRING GULL
Larus argentatus

race *argenteus*

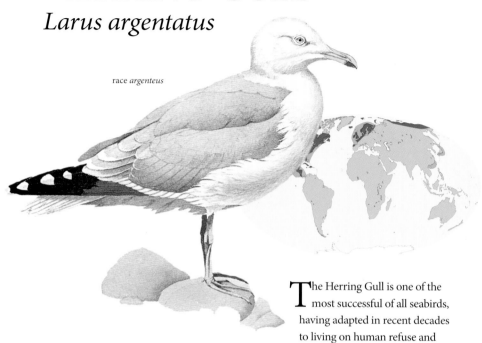

The Herring Gull is one of the most successful of all seabirds, having adapted in recent decades to living on human refuse and nesting on roofs in many areas. It is now commonplace in towns and some appear well inland—strenuous efforts have been made to prevent the gulls from nesting on, and fouling, important buildings. The nest is an untidy gathering of vegetation and scraps of garbage. Soon after they have hatched in the nest, the young Herring Gulls peck instinctively at the prominent red spot on the lower mandible of their parents' bills. This action induces the adult bird to regurgitate the food in its crop. The race *L. a. argentatus* breeds in northern Europe; the pale race *L. a. argenteus* in the north-west. Yellow-legged groups in the south have been separated as distinct species, including the Yellow-legged Gull *L. michahellis* in southern Europe.

FACT FILE

RANGE
Circumpolar N, temperate and Mediterranean

HABITAT
Varied; coastal and inland, including urban areas

SIZE
22–26 in (56–66 cm)

COMMON BARN-OWL
Tyto alba

race *alba*

In many areas, these birds nest in buildings, such as barns, outbuildings, and ruins, but in others, they prefer holes in trees. They hunt mostly on the wing, low down, with a light, buoyant action, frequent changes of direction and, sometimes, periods of hovering. They can locate and capture their prey solely by using their phenomenal hearing. Small rodents are normally the main prey and cyclic fluctuations in their numbers partly govern the numbers and breeding success of the owls. Most Barn Owls, such as the race *T. a. alba* from Britain, France, southern Europe, and Tenerife and Gran Canaria in the Canary Islands, are strikingly white below and pale sandy-gold and gray above. By contrast, other races, such as *T. a. guttata* from central and eastern Europe, have a dark buffish breast.

FACT FILE

RANGE
Americas, Europe, Africa, Arabia, India, Southeast Asia, Australia

HABITAT
Mainly open or semi-open lowlands, including farmland

SIZE
13–14 in (33–35 cm)

Eurasian Tawny Owl

Strix aluco

race *aluco*
gray phase

chick

race *sylvatica*
rufous phase

Seldom seen, due to its strictly nocturnal habits, the Eurasian Tawny Owl is nevertheless better known than other European owls through its familiar song—a long, quavering hooting, used to proclaim territorial ownership, advertise its presence to a mate, and used during courtship. The male's hoot is slightly lower pitched and more clearly phrased than that of the female; sometimes they perform a duet to strengthen the pair-bond. More often, however, the female replies with a *kewick* contact call. The Eurasian Tawny Owl is primarily a hole-nester, most often in trees and, like other owls, the female remains with the young while the male hunts. The birds feed mainly on small mammals, but they are very adaptable and will readily switch to other prey, such as small birds. This versatility allows a wide range. Both reddish and grayish forms are found in most areas; the grayer birds are most common in the north and east."

FACT FILE

RANGE
Europe, N Africa, parts of W Asia, China, Korea, Taiwan

HABITAT
Chiefly woodland or areas with some trees, including urban areas

SIZE
14½–15 in (37–39 cm)

EURASIAN SCOPS OWL
Otus scops

brown
individual

gray
individual

This small, superbly camouflaged owl (which may be either gray-brown or red-brown in basic color) is particularly hard to detect at its daytime roost. Indeed, it is far more often heard than seen and its song—a monotonously repeated single, plaintive, whistling, note—being one of the most characteristic sounds of the Mediterranean night. The Scops Owl is unusual among European owls in being largely a true summer migrant. Large insects, such as crickets, cicadas, cockchafers, and moths are its main prey, with moths frequently taken in flight and often attracting the owls to hunt around streetlights.

FACT FILE

RANGE
Breeds S Europe, N Africa, E to SW Siberia; N populations and some S ones, winters in tropical Africa

HABITAT
Open woodland, parks, orchards, gardens, roadside trees, town squares

SIZE
7½–8 in (19–20 cm)

WOOD PIGEON
Columba palumbus

FACT FILE

RANGE
Europe, N Africa, SW Asia, Iran and India; also Azores

HABITAT
Woodland and adjacent farmland

SIZE
16–16½ in (40–42 cm)

Wood Pigeons are serious pests of agricultural crops in some areas, with flocks sometimes numbering thousands or even tens of thousands. These birds frequently raid fields planted with cereals, brassicas (cabbages, kale, and brussels sprouts), clover, and peas. The distinctive display flight of the male Wood Pigeon (also rarely performed by the female) consists of a towering flight with several slow wing beats toward the top, when the wings produce 2–3 very loud claps. These sounds may come either from direct contact between the wings or by a whip-crack effect of the feather tips. The bird then glides downward and flies up again to repeat the process.

ROCK DOVE
Columba livia

The wild Rock Dove is a bird of cliffs and rocky terrain but its feral descendants thrive in towns and cities throughout the world. Originally, wild-caught Rock Doves were reared for food and, later, for producing fancy breeds and racing pigeons. Escapes, which still continue, soon led to feral populations gathering in urban areas, using building ledges for nest sites and gleaning scraps of food from the streets and surrounding fields. These Ferel Pigeons vary greatly in plumage. While some are similar to Rock Doves in coloration, others are rusty red or checkered. The remarkable homing ability exploited in racing pigeons is at odds with the almost entirely sedentary habits of both wild and feral birds. In the wild, movements of more than a few miles are rare, yet racing pigeons can find their way home from distances of hundreds or even thousands of miles without difficulty.

FACT FILE

RANGE
Native to S Eurasia and
N Africa; domesticated
worldwide

HABITAT
Cliffs and gorges, close to
open country; domestic form
mainly urban

SIZE
12–13 in (31–34 cm)

MOURNING DOVE
Zenaida macroura

This long-tailed sandy dove shows white outer tail feathers in flight. Outside of the breeding season, Mourning Doves are generally found in flocks and roost communally. They are mostly ground feeders, taking a variety of grain and seeds and, occasionally, small invertebrates. The advertising call is a series of 3–4 mournful coos, from which the birds take their common name. They produce 2 or 3 broods each season and have benefited from the conversion of forests to farmland. In the deep South, they begin nesting in February and continue until the end of October, sometimes having as many as 6 nesting attempts in a season. They are monogamous and have elaborate courtship displays, with the male strutting before the female or chasing her while puffing up his feathers and cooing. The males also perform display flights with periods of noisy flapping and gliding. They are solitary nesters and build a flimsy platform in shrubs or trees.

FACT FILE

RANGE
SE Alaska, S Canada to C Panama, Caribbean

HABITAT
Farms, open woods, towns, gardens

SIZE
12 in (30 cm)

CHIMNEY SWIFT
Chaetura pelagica

nest

This small, dark swift is an extremely gregarious bird, which nests and roosts in huge colonies, sometimes numbering several thousand birds, which may overlap like shingles on a roof. Originally, the nest sites and roosting sites of the Chimney Swift were hollow trees but today the colonies take over air shafts or large chimneys. The nest is made of thin twigs, which the Chimney Swift snatches in its feet while in flight. There is some evidence of cooperative breeding, with "helpers" assisting the parents with rearing the young.

FACT FILE

RANGE
Breeds E U.S.A. and S Canada; winters in Central and NE South America

HABITAT
Open and wooded areas, often in towns

SIZE
5 in (13 cm)

EURASIAN SWIFT
Apus apus

No land bird is more aerial than the Eurasian Swift. It feeds, drinks, sleeps, and mates on the wing, returning to earth only to lay its eggs and feed its chicks. A young bird may spend up to 4 years in the air, from the moment it leaves the nest to the day it returns to the colony to rear its own family. Not surprisingly, it is a superb flier—wheeling, climbing, and diving as it scoops flying insects from the air, or engages in excited chases, uttering its characteristic shrill, screaming calls. These swifts nest in colonies, mostly in roof spaces. The adults feed their young on insects carried back in their throat pouches. When insects are scarce during cold or adverse weather, the young will survive for several days without food by becoming torpid, so reducing energy loss.

FACT FILE

RANGE
Most of Europe, parts of N Africa and C Asia and E almost to pacific; winters in tropical Africa

HABITAT
Aerial; breeds in buildings and rock crevices

SIZE
5½ in (14 cm)

AFRICAN PALM SWIFT
Cypsiurus parvus

nest

This slender, pale brown swift is always found in association with palms and may gather in huge flocks at favorite trees. Even in the middle of urban areas it may be very common indeed. It usually forages in groups at treetop height, circling and jinking on its long, narrow wings and occasionally spreading its deeply forked tail as it banks into a steep turn. Its nest is little more than a pad of feathers and plant down, glued to the vertical underside of a palm frond with saliva. The 2 eggs are also glued to the nest and the parents incubate them in turns, clinging vertically to the nest. The nestlings have long curved claws, enabling them to cling on tightly as the palm fronds are whipped about violently during windy weather.

FACT FILE

RANGE
Sub-Saharan Africa, except much of Ethiopia, Somalia, and S Africa

HABITAT
Aerial, near plants

SIZE
6 in (15 cm)

EASTERN BLUEBIRD
Sialia sialis

The Eastern Bluebird often forms flocks in the fall and winter and several birds may roost together in the same cavity during winter. Typically, it has a hunched appearance when perching. It often flies down from its perch to catch insects on the ground. In colder months, berries become an important part of the diet. Nesting usually takes place in natural tree cavities and old woodpecker holes. Although the eggs of this species are typically blue, some individual females lay clutches of white eggs. Multiple broods are common, with 2 chicks often produced in a single season in northern areas, and up to 4 in southeastern U.S.A. In many areas, the Eastern Bluebird has declined in numbers because of the competition for nest sites with other birds. The provision of nest-boxes by conservation organizations and concerned individuals has helped the species.

FACT FILE

RANGE
E North America and Central America

HABITAT
Open woodland, roadsides, farms, orchards, gardens, parks

SIZE
5½–7½ in (14–19 cm)

BLACK-BILLED MAGPIE
Pica pica

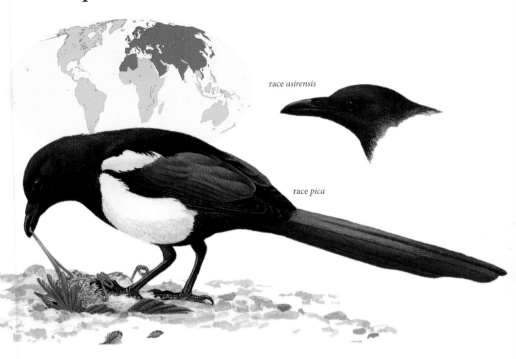

race *asirensis*

race *pica*

The Black-billed Magpie has a bad reputation for plundering the nests of other birds and it is often shot on sight on estates where there are gamebirds. Elsewhere it flourishes, for it is an adaptable bird with a broad diet. It has adapted well to the man-made landscape and is now a familiar sight in suburban gardens. Most races are similar to the race *P. p. pica* of the British Isles, central and eastern Europe though the race *P. p. asirensis* of western Arabia is distinctive and is very dark with less white, and a large bill.

FACT FILE

RANGE
Eurasia, from W Europe to Japan; temperate North America

HABITAT
Open woodland, scrub, farmland, towns

SIZE
18 in (45 cm)

WESTERN JACKDAW
Corvus monedula

juv

FACT FILE

RANGE
W Europe, excluding extreme N, to C U.S.S.R.

HABITAT
Woods, parks, gardens, farmland, sea cliffs, towns

SIZE
13 in (33 cm)

The Western Jackdaw is the smallest of the *Corvus* crows in Europe and one of the most familiar. This is partly because it is a hole-nesting bird that has learned to use the cavities in church towers, chimney stacks, and other parts of tall buildings as artificial substitutes for tree holes and rock crevices. It is also common on sea and inland cliffs and quarries, where it still breeds in more natural surroundings. In such places, its agility in the air can be seen to great advantage. Western Jackdaws enjoy a broad diet, which may include anything from caterpillars to crabs. They also scavenge for refuse and rob the nests of other birds. In spring, they collect soft fiber for nesting material and are often seen plucking wool from the backs of sheep.

EASTERN PHOEBE
Sayornis phoebe

Named for its pleasantly raspy fee-bee song, which is delivered incessantly during the breeding season, the Eastern Phoebe is a familiar sight on farmland and wooded roadsides throughout much of North America. It hunts in typical flycatcher fashion, sallying after insects from an exposed perch, and wagging its tail from side to side in a distinctive manner as it alights. When prospecting above a quiet pond surface for flies it may hover briefly or flutter in pursuit like a butterfly, its bill clicking as it snaps up a meal. An opportunist nester, it often uses man-made structures as breeding sites, favoring niches protected from the weather, such as the rafters of farm outbuildings, porches, and bridge supports.

FACT FILE

RANGE
Breeds N and E North America; winters SE and SC U.S.A. to Mexico

HABITAT
Woods, farms, usually near water

SIZE
7 in (18 cm)

BARN SWALLOW
Hirundo rustica

race *erythrogaster*

race *rustica*

FACT FILE

RANGE
Breeds Europe and Asia,
N Africa, North America;
winters in S hemisphere

HABITAT
Open areas,
often near water

SIZE
7 in (18 cm)

Although almost cosmopolitan, this species only has 6 races, including the widespread *H. r. rustica*, which breeds over much of Eurasia and north Africa, and *H. r. erythrogaster* of North America. The Barn Swallow has benefited greatly from the destruction of forest habitats, the construction of buildings and bridges, and the provision of utility wires and fences. Over much of its nesting range, it attaches its mud nest to the rough walls of barns and other human structures and depends on wires for use as perches. Barn Swallows prey on small flying insects that they catch mainly during low flights over water or moist vegetation.

HOUSE MARTIN
Delichon urbica

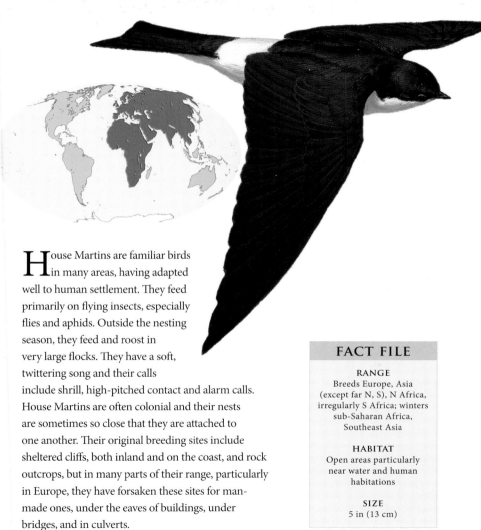

House Martins are familiar birds in many areas, having adapted well to human settlement. They feed primarily on flying insects, especially flies and aphids. Outside the nesting season, they feed and roost in very large flocks. They have a soft, twittering song and their calls include shrill, high-pitched contact and alarm calls. House Martins are often colonial and their nests are sometimes so close that they are attached to one another. Their original breeding sites include sheltered cliffs, both inland and on the coast, and rock outcrops, but in many parts of their range, particularly in Europe, they have forsaken these sites for man-made ones, under the eaves of buildings, under bridges, and in culverts.

FACT FILE

RANGE
Breeds Europe, Asia
(except far N, S), N Africa,
irregularly S Africa; winters
sub-Saharan Africa,
Southeast Asia

HABITAT
Open areas particularly
near water and human
habitations

SIZE
5 in (13 cm)

PURPLE MARTIN
Progne subis

♂

FACT FILE

RANGE
Breeds North America;
winters in Amazon Basin,
and sometimes as far
N as Florida

HABITAT
Open areas, often near water

SIZE
7 in (18 cm)

This species has become intimately associated with human settlement in eastern North America. The Indians of southeastern U.S.A. provided calabash gourds for Purple Martins to use as nest sites. In return, the birds mobbed crows and other animals that fed on the Indians' crops or on meat hung out to dry. Early European settlers continued the tradition of providing homes for the birds. In western North America, the birds still nest in natural sites, usually dead trees with woodpecker holes. Both parents feed the young, which fledge within about 4 weeks of hatching. Young birds can catch flying insects within 5 days of fledging and are quickly independent of their parents.

Pied Wagtail
Motacilla alba

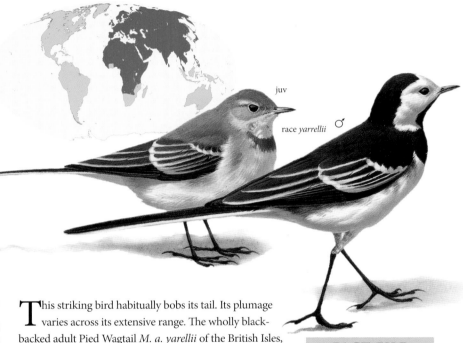

juv

race *yarrellii* ♂

This striking bird habitually bobs its tail. Its plumage varies across its extensive range. The wholly black-backed adult Pied Wagtail *M. a. yarellii* of the British Isles, Spain, and Morocco differs markedly from the gray-backed White Wagtail *M. a. alba* of the rest of Europe, north and east Africa, Iran, Arabia, and Russia. The Pied Wagtail is fond of water and often paddles into shallows and puddles as it searches for aquatic insects. It is also common in open ground away from water, and regularly occurs around farmyards, parks, and buildings. It often uses communal roosts outside of the breeding season. In some areas, these include artificially heated structures, such as power station cooling towers and heated greenhouses.

FACT FILE

RANGE
Breeds most of Eurasia, except tropics; N birds winter S to Africa N of Equator, Arabia, India, and Southeast Asia

HABITAT
Open ground, often near water

SIZE
7 in (18 cm)

WINTER (NORTHERN) WREN
Troglodytes troglodytes

race *zetlandicus*

The tiny, cock-tailed Winter Wren, the only wren that occurs in the Old World, is equally at home in a damp, broad-leaved wood, in the gorse and bracken of an upland moor, or on the cliffs of a windswept Atlantic or Pacific island. There are nearly 40 races, including *T. t. indigenus* of the British mainland, *T. t. zetlandicus* of the Shetland Islands and *T. t. helleri* of Kodiak Island, off Alaska. Races are paler in hot, desert regions and darker in wetter places.

FACT FILE

RANGE
Breeds C and S Canada, Alaska, coastal W U.S.A., parts of E U.S.A., Europe, Asia E to Japan, N Africa; N populations winter to S

HABITAT
Woodland, gardens, cultivated land, moors, heaths, rocky islands

SIZE
3 in (8 cm)

race *indigenus*

RED-LEGGED THRUSH
Turdus plumbeus

This striking bird is the most widespread of the West Indian thrushes. Its plumage varies among the islands, the chin and throat ranging from white to black, and the belly from reddish-yellow to gray and white. Its song is rather weak and melancholy and, when disturbed, the bird utters a loud *wet-wet* call, sometimes repeatedly. The Red-legged Thrush frequents a variety of habitats. In some areas, it is shy and difficult to observe, while in others, it is quite bold and easy to watch. It commonly forages on the ground, making loud, rustling sounds as it searches under dead leaves for insects to eat. It has an omnivorous diet, including insects, reptiles, and fruit. The cup nest of the Red-legged Thrush is built in a bush, in the fork of a branch or among the fronds of a palm.

FACT FILE

RANGE
West Indies

HABITAT
Forest areas in mountains and lowlands, gardens, city lawns

SIZE
10–11 in (25–28 cm)

NORTHERN MOCKINGBIRD

Mimus polyglottos

juv

FACT FILE

RANGE
S Canada, U.S.A., Mexico, Caribbean islands; introduced to Bermuda, Hawaiian islands

HABITAT
Open areas in cities, suburbs, countryside, deserts

SIZE
9–11 in (23–28 cm)

This familiar bird is the state bird for 5 U.S.A states. It is an accomplished mimic, usually repeating learned phrases in each song. Its diet includes fruit, berries, and invertebrates, and good sources of food are often vigorously defended against all competing species. This bird's habit of wing flashing—pausing as it runs along the ground and deliberately raising and partially opening its wings—is a common but, as yet, poorly explained form of behavior. It may be a way of startling potential prey out of hiding. Breeding begins as early as mid-February. Nests are well concealed in bushes or trees, made of sticks and lined with grass, leaves, and plant down. Pieces of string, cotton, and plastic are often woven into them.

DUNNOCK
Prunella modularis

This generally unobtrusive little bird spends much of its time in cover among shrubs and hedgerows, but also forages on the open ground for its food with a curious, shuffling, jerky, rather mouselike gait; it can also hop. In winter, it feeds mainly on seeds and in summer eats chiefly insects and their larvae. The bird is generally known in Britain by its old English name Dunnock, from its gray-brown, or dun, coloration. Dunnocks have 3 different types of territory: those held by a solitary male, those held by a conventional male-female pair, and larger territories held by a male-female pair plus an additional male. Often one male associates with 2 or 3 females, or 2 or 3 males with up to 4 females.

FACT FILE

RANGE
Europe S to C Spain and Italy and E to the Urals, Lebanon, Turkey, N Iran, Caucasus

HABITAT
Scrubs, heathland, mixed woodland, young coniferous forest, farmland hedgerows, parks, gardens, vacant urban land, scrubby coastal cliffs and dunes

SIZE
5½ in (14 cm)

AUSTRALIAN MAGPIE
Gymnorthina tibicen

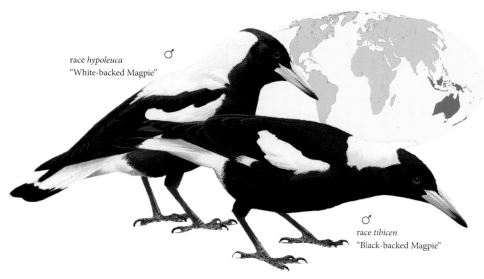

race *hypoleuca*
"White-backed Magpie" ♂

♂ race *tibicen*
"Black-backed Magpie"

The plumage on the back of the Australian Magpie varies according to race. For example, the race *G. t. tibicen*, often called the Black-backed Magpie, from central and northern Australia differs markedly from the race *G. t. hypoleuca*, often called the White-backed Magpie, from southern Australia and Tasmania. Australian Magpies eat scarab beetle larvae and other invertebrates, listening for them in the soil and then digging them up with their strong, pointed bills. These birds have a complex social organization, linked to habitat. Permanent groups, with a high breeding success, live in prime habitat with trees and grassland. More transitory groups live in poorer quality sites and non-breeding flocks may occupy the areas of grassland with the fewest trees.

FACT FILE

RANGE
Australia, S New Guinea;
introduced to New Zealand

HABITAT
Woodland and grassland

SIZE
16 in (40 cm)

WILLIE WAGTAIL
Rhipidura leucophrys

The Willie Wagtail is one of the most common and popular Australian birds. This is due to its striking plumage, its tameness, its pugnacity toward other birds, and its conspicuous feeding habits. Wagtails spend half their time darting after flies, beetles, and ants on the ground. They wag and twist their tails from side-to-side, often flicking their wings. At other times, they fly out after insects from low branches, stumps, or even from a sheep's back. During the breeding season, they are very aggressive, attacking potential predators up to the size of an eagle, vigorously fluttering around the intruder's back and head with persistent chattering calls. The nest is conspicuous, but, perhaps because the birds are so aggressive, 65 percent of nests successfully produce young."

FACT FILE

RANGE
Throughout Australia; rare in Tasmania, New Guinea, Soloman Islands, Bismarck Archipelago and Moluccas

HABITAT
Open woodland, scrub, grassland

SIZE
20 in (8 cm)

**Northern
Bullfinch**
Pyrrhula pyrrhula
This bird feeds on
seeds and the buds of
fruit trees.

Common Waxbill
Estrilda astrild
Grass seeds are
favorites for
this bird.

Pine Grosbeak
Pinicola enucleator
The Pine Grosbeak
feeds on seeds, berries,
and buds.

Seed-eating Birds

Seed-eating birds are typical garden visitors. They have conical beaks with sharp cutting edges for peeling off husks, and strong muscles for cracking seeds. Finches, buntings, and sparrows have evolved a variety of bills that are perfectly suited to specific tasks. Bullfinches and pine grosbeaks have wide, rounded bills that are ideal for eating soft seeds and buds; goldfinches have longer beaks for tackling spiny seed heads; greenfinches and evening grosbeaks have massive bills that deal with larger, harder seeds and berries.

Even seed-eaters, however, vary their diet and need insect food in spring and summer. Nevertheless, they are some of the easiest birds to attract into a garden.

Some seed-eaters eat their food on the bird table or feeder itself, while others take seeds away to eat. Some of the titmice and chickadees, and members of the crow family, carry seeds and nuts away to hide elsewhere. They return to this cache of food later when they have more time to feed, or perhaps when food is in short supply.

Village Weaver
Ploceus cucullatus
This bird feeds on grain and seeds.

Dickcissel
Spiza americana
Weed seeds are one of the Dickcissel's main foods.

WHITE-THROATED SPARROW

Zonotrichia albicollis

The White-throated Sparrow is familiar in much of North America east of the Rocky Mountains and is a common visitor to backyard feeders, being especially fond of millet and oil-rich black sunflower seeds. There are white-crowned and tan-crowned forms. Males of both types prefer females with white stripes, but both kinds of females prefer tan-striped males. White-striped birds appear to be more aggressive than tan-striped ones. White-throated Sparrows forage on the ground in dead leaves, often in flocks, in woods, forest edges and clearings, and in coppice near the treeline. In winter, they come to thickets, overgrown fields, parks, and suburbs. They eat seeds and fruit, but in summer, they also feed on large numbers of insects, spiders, and other small invertebrates. They nest on or near the ground with the nest well concealed by leaves.

CHIPPING SPARROW
Spizella passerina

The Chipping Sparrow is a tame, confiding species that is well-known to people who provide bird-feeding stations. It eats large quantities of grass and weed seeds, as well as weevils, leafhoppers, and other insects. On occasion, it will dart out to catch flying insects. Its song is a series of "chips," which at times, the bird runs together in a single-pitched, rapid trill. The Chipping Sparrow sometimes sings at night. In winter, the adult's bright chestnut crown becomes more dull and streaked, its prominent white eyebrow is lost and it acquires a brown ear patch. The female carries out all the nest-building duties, although the male may accompany her while she is gathering material.

FACT FILE

RANGE
Canada S to N Nicaragua

HABITAT
Gardens, woodland, forest edge in E; open woodland and mountains in W

SIZE
5–5½ in (12.5–14 cm)

SONG SPARROW
Melospiza melodia

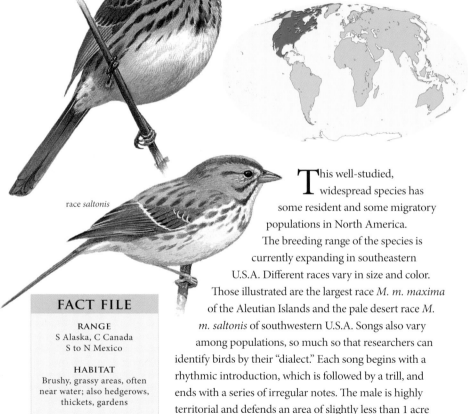

race *maxima*

race *saltonis*

This well-studied, widespread species has some resident and some migratory populations in North America. The breeding range of the species is currently expanding in southeastern U.S.A. Different races vary in size and color. Those illustrated are the largest race *M. m. maxima* of the Aleutian Islands and the pale desert race *M. m. saltonis* of southwestern U.S.A. Songs also vary among populations, so much so that researchers can identify birds by their "dialect." Each song begins with a rhythmic introduction, which is followed by a trill, and ends with a series of irregular notes. The male is highly territorial and defends an area of slightly less than 1 acre (0.4 hectares) in prime habitat. The Song Sparrow eats mainly seeds, but also eats insects and other invertebrates.

FACT FILE

RANGE
S Alaska, C Canada
S to N Mexico

HABITAT
Brushy, grassy areas, often near water; also hedgerows, thickets, gardens

SIZE
5½ in (14 cm)

DICKCISSEL
Spiza americana

♂

The common name of this little bird is derived from its constantly repeated call, which sounds like "dick-dick cissel." The male is characterized by his heavy bill, yellow breast, black bib, and chestnut wing patches. The female is sparrowlike, with streaky plumage, but has a yellowish breast and small chestnut wing patches. Dickcissels feed mainly on the ground on weed seeds, grain, spiders, and insects, such as grasshoppers and crickets. The male Dickcissel establishes the nesting territory by calling from a conspicuous perch. The cup-shaped nest, made of weed stems, grasses, and leaves, is built on the ground, often in the shelter of alfalfa or grain crops, or in a tree or bush. The female usually lays 4 eggs, which she incubates for 12–13 days. There are generally 2 broods a year.

FACT FILE

RANGE
Great Lakes, S through
Midwest to Gulf states;
winters from Mexico to
N South America

HABITAT
Grassland, grain,
and alfalfa fields

SIZE
6–7 in (15–18 cm)

REED BUNTING
Emberiza schoeniclus

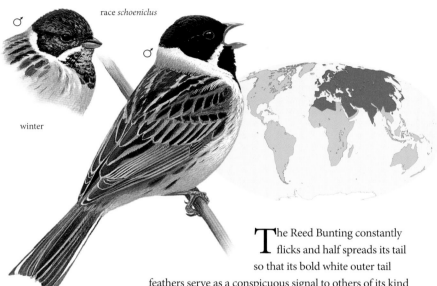

race *schoeniclus*

♂

♂

winter

The Reed Bunting constantly flicks and half spreads its tail so that its bold white outer tail feathers serve as a conspicuous signal to others of its kind in the dense reedbeds and marshes. With the drainage of wetlands, this attractive bunting declined over much of its range, but during the past 50 years or so it has shown an increasing tendency to colonize drier habitats, such as scrubland, overgrown ditches, and hedgerows. The shape of the Reed Bunting's bill varies geographically to a striking extent, from very small, slight, and pointed in races, such as *E. s. parvirostris* of central Siberia to thick and almost parrot-shaped in races, such as *E. s. pyrrhuloides* of western and central Asia, western Mongolia, and northwest China. The race *E. s. schoeniclus* from northwestern Europe and central Russia is one of the smaller-billed races, although its bill is stouter than that of *E. s. parvirostris*.

FACT FILE

RANGE
Eurasia S to Iberia and S U.S.S.R. and E to NE Asia and Manchuria; N and E populations winter S as far as N Africa, Iran, and Japan

HABITAT
Lowland marshes, reedbeds and, increasingly, drier habitats

SIZE
6–6½ in (15.5–16.5 cm)

PINE GROSBEAK
Pinicola enucleator

These large finches are often elusive in their northern forest homeland. They are scarce and unobtrusive. Adult males are brightly colored, but young birds and females have somber hues. Pine Grosbeaks eat mainly berries and buds, hopping about heavily on the ground, or clambering around slowly in the treetops. While many birds are sedentary, others leave the breeding grounds and travel south and west in winter, sometimes in large numbers. The magnitude of these movements is related to the abundance of the Pine Grosbeak's natural food; in Europe, the size of the crop of rowan berries is especially important. Few birds are seen in years with poor crops. During winter, Pine Grosbeaks are often seen in small flocks, when they may be remarkably tame and approachable.

FACT FILE

RANGE
N and W North America,
N Scandinavia E to N Siberia

HABITAT
Coniferous and scrub forest

SIZE
8 in (20 cm)

Northern Bullfinch
Pyrrhula pyrrhula

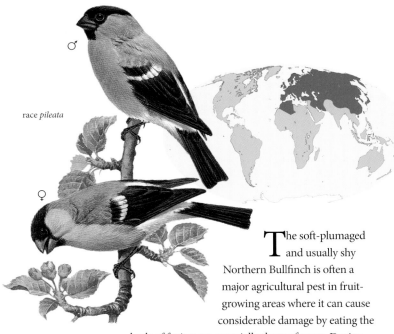

race *pileata*

♂

♀

FACT FILE

RANGE
Europe E across
Asia to Japan

HABITAT
Coniferous and broad-
leaved forest, cultivated
areas, gardens

SIZE
6 in (15 cm)

The soft-plumaged and usually shy Northern Bullfinch is often a major agricultural pest in fruit-growing areas where it can cause considerable damage by eating the buds of fruit trees, especially those of pears. During the breeding season, it carries food to the young in special pouches on the floor of its mouth, either side of the tongue. This is in contrast to most finches, which carry food in their throats. At the eastern end of the Northern Bullfinch's range, in the Amur region of the U.S.S.R., Manchuria, and Japan, a race *P. p. griseiventris* has a gray breast and red face. In the Azores, the local bullfinch is a separate, endangered species, the Azores Bullfinch *P. murina*, in which the sexes are alike, lacking the pink breast and the white rump shared by all Northern Bullfinch races.

Red-billed Fire Finch
Lagonosticta senegalia

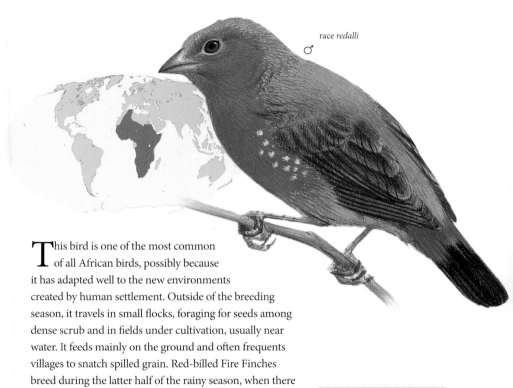

race *redalli*
♂

This bird is one of the most common of all African birds, possibly because it has adapted well to the new environments created by human settlement. Outside of the breeding season, it travels in small flocks, foraging for seeds among dense scrub and in fields under cultivation, usually near water. It feeds mainly on the ground and often frequents villages to snatch spilled grain. Red-billed Fire Finches breed during the latter half of the rainy season, when there are plenty of ripe seeds to feed to the nestlings. In wild country, the monogamous pairs nest in shrubs, particularly acacias, but in settled areas they will nest in thatched roofs or wall cavities. The nests are often parasitized by a parasitic viduine weaver, the Village Indigobird *Vidua chalybeata*. There are 6 or more races recognized, including *L. s. rendalli* from southern Angola to southern Tanzania and the northern Cape province.

FACT FILE

RANGE
Sub-Saharan Africa

HABITAT
Dry areas, dense brush, and near cultivated land

SIZE
3½ in (9 cm)

COMMON WAXBILL
Estrilda astrild

race *astrild*

The Common Waxbill feeds mainly on grass seeds, either plucking them from the seed heads or picking them off the ground. It will also prey on swarming termites. Highly gregarious, these birds often forage in large flocks and return to communal roosts at night. They breed in the rainy season, when insects for feeding their young are most abundant. The nest is often highly elaborate, with a small "roosting" nest built on top of the dome-shaped breeding nest. The brood is often parasitized by the Pin-tailed Whydah. In contrast to the chicks of some parasitic cuckoos, such as the Eurasian Cuckoo *Cuculus canorus*, those of the whydah do not destroy the rightful occupants of the nest, and the adult waxbills can often be seen feeding a mixed brood. There are some 8 races, including *E. a. astrild* from much of South Africa.

FACT FILE

RANGE
Sub-Saharan Africa;
introduced to many
tropical islands

HABITAT
Grassland and
cultivated areas

SIZE
4 in (10 cm)

GROSBEAK WEAVER
Amblyospiza albifrons

race *albifrons*

An alternative name for this large, gregarious species is Thick-billed Weaver. The striking male Grosbeak Weaver reveals white patches in the wings when he flies with a rather undulating flight. The female Grosbeak Weaver shares the distinctive flight but is a much more drab-looking bird than the male and is brown above and whitish with dark streaks below. There are several races, including *A. a. albifrons*, illustrated, from the Cape Province. The male weaves the beautifully constructed sphere-shaped nest from thin grass strips and fixes it to 2 or 3 upright grass stems. The entrance to the nest is a small hole in the side, with or without a "porch" depending on the location, and nests are often in small colonies.

FACT FILE

RANGE
Much of sub-Saharan Africa

HABITAT
Swampy woodland,
damp forest, reedbeds

SIZE
6½–7½ in (17–19 cm)

VILLAGE WEAVER
Ploceus cucullatus

♂ race *collaris*

nest

♂

race *spilonotus*

FACT FILE

RANGE
Sudan and Ethiopia S to
Angola, Cape Province

HABITAT
Forest, cultivated
land, gardens

SIZE
6–7 in (15–18 cm)

The males of this widespread species have a strikingly patterned plumage in the breeding season, but females and non-breeding males have greener, streaked plumage. There are some 7 races, including *P. c. collaris* of Zaire and N Angola and *P. c. spilonotus* of South Africa, which vary according to the pattern of black on the head and the strength of coloring below. Village Weavers, also called Spotted-backed Weavers, breed in noisy, lively, dense colonies. Nests of grass strips woven into kidney-shaped balls, with or without a "spout," are slung from the tips of spreading, drooping branches, often over water. The male Village Weavers build the nests.

GOLDEN PALM WEAVER
Ploceus bojeri

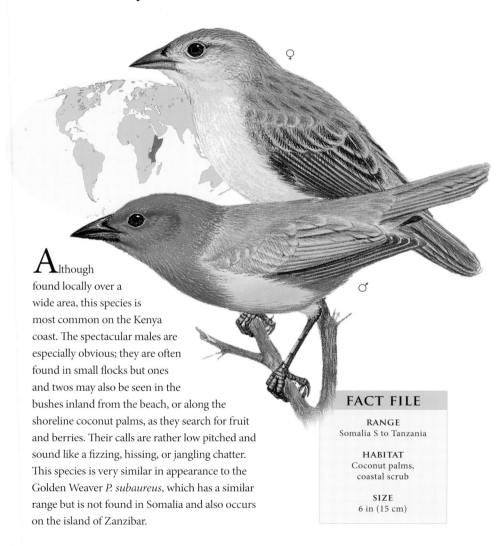

♀

♂

Although
found locally over a
wide area, this species is
most common on the Kenya
coast. The spectacular males are
especially obvious; they are often
found in small flocks but ones
and twos may also be seen in the
bushes inland from the beach, or along the
shoreline coconut palms, as they search for fruit
and berries. Their calls are rather low pitched and
sound like a fizzing, hissing, or jangling chatter.
This species is very similar in appearance to the
Golden Weaver *P. subaureus*, which has a similar
range but is not found in Somalia and also occurs
on the island of Zanzibar.

FACT FILE

RANGE
Somalia S to Tanzania

HABITAT
Coconut palms,
coastal scrub

SIZE
6 in (15 cm)

Goldcrest
Regulus regulus
The Goldcrest feeds
mainly on small insects.

American Redstart
Setophaga ruticilla
This bird preys
on caterpillars.

Cardinal Honeyeater
Myzomela cardinalis
This bird will sip nectar
from flowers.

Noisy Friarbird
Philemon corniculatus
This bird has a sugar-
rich diet of fruit, nectar,
as well as insects.

NECTAR AND INSECT FEEDERS

Common Tody Flycatcher
Todirostrum cinereum
This bird forages
for insects.

Tropical gardens are rich in large, colorful flowers, and birds are quick to exploit the opportunities they offer. Many produce prodigious quantities of nectar, specially to attract insects that incidentally pollinate the flowers. Birds eat both the nectar and the insects that visit the blooms. Some, such as hummingbirds, specialize in the high-energy nectar; others, such as sunbirds and honeyeaters, eat both nectar and insects.

Red-whiskered Bulbul
Pycnonotus jocosus
Ants form part of
this bird's diet.

Hummingbirds can also be tempted by feeders supplying sugar-rich solutions. Sunbirds and hummingbirds may have long bills, adapted both in length and shape to take nectar from specific species of wildflowers. Some sunbirds, however, have shorter bills and peck through the base of the petals to literally take a shortcut to the nectar, damaging the flower in the process.

In more temperate regions, some species feed on nectar or pollen, but only as an alternative to more regular foods. However, many catch insects that are attracted to flowers, including late-flowering species, such as ivy in early fall.

SOUTHERN BLACK TIT
Parus niger

FACT FILE

RANGE
SE Africa, N to Tanzania

HABITAT
Woods, woodland edge,
lightly wooded country

SIZE
6 in (16 cm)

Found in a wide variety of woodland, ranging from dense evergreen forest to scattered trees, the Southern Black Tit breeds in family groups of 3–4 birds—the "helpers" are generally the breeding pair's male offspring from the previous year's brood. But, despite this extra food-gathering capacity, each family raises only 3–4 chicks each season. Though this species is restricted to the east of southern Africa, it is very similar to a bird known as Carp's Tit, which is found in some parts of the west and currently held to be a separate species, *P. carpi*. The Southern Black Tit is very similar to the closely-related Great Tit *Parus major* of Europe and Asia, in general form and behavior.

CHIN-SPOT PUFF-BACK FLYCATCHER
Batis molitor

The Chin-spot Puff-back Flycatcher is a restless bird, hawking after insects like a true flycatcher and hovering in the air while searching through leaves. It will also forage in lower branches like a tit. Its wings produce a rattle in flight. The bird builds a small, cup-shaped nest of fibers and lichens, lashed together with spiders' webs. It is usually well concealed in the fork of a tree. The 2 eggs are grayish or greenish white with a girdle of dark spots.

FACT FILE

RANGE
S Sudan, Kenya
(except coast), SW Africa,
Mozambique and
E Cape Province

HABITAT
Open acacia woodland,
forests and their edges,
cultivated areas, gardens

SIZE
4½ in (11.5 cm)

BLUE FLYCATCHER
Elminia longicauda

imm

♂

FACT FILE

RANGE
W Africa, E to Kenya,
S to Angola

HABITAT
Woodland and forest,
gardens, farmland

SIZE
5½ in (14 cm)

Beautifully colored in pale cerulean blue and equally elegant in shape, the Blue Flycatcher is one of the most attractive of African flycatchers. It is typically alert, upright, and full of nervous energy, frequently fanning its rather long, graduated tail. At the same time, though it is always active and ready to chase some flying insect, it is very tame and confiding. It has a brief, twittering song, which is somewhat insignificant but a little reminiscent of the songs of several of the sunbirds. Blue Flycatchers live in small groups that may contain 2 or 3 pairs or trios of adults with immatures or dependent young. A leader defends the territory. Females lay 1 or 2 white eggs.

AFRICAN PARADISE FLYCATCHER

Terpsiphone viridis

chestnut phase

♂ ♀

In shady, mixed forest, or beside rivers and lakes where woodland glades and ornamental gardens provide relief from the dense canopy, the gorgeous African Paradise Flycatcher flits out from a hidden perch to catch an insect in midair, giving a sudden flash of color and a glimpse of a long, trailing tail. Its call is a distinctive, sharp double or treble note. While the tails of males may be up to 8 in (20 cm) long, females usually have short tails. However, tail length, whether long or short, is not always an accurate guide to the sex of the bird. The nest is a splendidly neat, surprisingly small construction of fibers and fine roots, bound and camouflaged with spiders' webs and lichens. It is always on a slender, open, bare twig in a shady spot, often over water, and the tail of the sitting bird trails elegantly over the side.

FACT FILE

RANGE
Sub-Saharan Africa

HABITAT
Forest and riverside woodland with mature trees

SIZE
16 in (40 cm)

BLUE-GRAY GNATCATCHER
Polioptila caerulea

♂

♀

The Blue-gray Gnatcatcher is typical of the gnatcatcher group; bluish-gray, slender, with a long tail, which it often cocks like a wren, and a long, thin, pointed bill. It is a tiny, restless bird, often pugnacious. It forages among the leaves and twigs of trees for insects. It will sometimes catch these on the wing, or pick them off leaves while hovering like a kinglet. Its song is a wheezy whisper. The nest is usually placed in a tree fork or on a branch. It is a beautifully constructed cup made of plant down, bound together with spider and insect silk and covered with pieces of lichen. Both adults build the nest and they also share the incubation of the 4–5 tiny, pale blue, brown-spotted eggs.

FACT FILE

RANGE
North America from S Canada to Guatemala and Cuba; winters S of South Carolina along Atlantic coast, and S of S Mississippi and S Texas

HABITAT
Forest, timbered swamps, thorny chaparral, wooded areas of towns

SIZE
4½–5 in (11.5–13 cm)

RUBY-CROWNED KINGLET

Regulus calendula

♂

Similar to the Goldcrest of Eurasia, this tiny, short-tailed American species is an active, nervous bird, which habitually flits its wings with sudden, jerking movements. This is a useful recognition point, since the characteristic scarlet patch on the crown is often concealed. It hunts assiduously for insects among the twigs and leaves and may also dart after any that fly past. It will also feed on fruit and seeds. In contrast with that of the other kinglets, such as the Goldcrest, the song of the male Ruby-crowned Kinglet is surprisingly rich. The nest is a distinctive cup-shaped construction of plant down, slung beneath a pine branch and covered in lichens and lined with feathers.

FACT FILE

RANGE
North America, from NW Alaska S to Arizona, also E Canada to Nova Scotia; winters S to N Mexico

HABITAT
Mixed woods, spruce bogs, fir woods

SIZE
4–4½ in (9.5–11.5 cm)

GOLDCREST
Regulus regulus

♂

♀

FACT FILE

RANGE
Discontinuous,
from Azores and NW
Europe and Scandinavia
to E Asia

HABITAT
Conifer woods to
14,700 ft (4,500 m), also
some broad-leaved woods

SIZE
3½ in (9 cm)

This attractive little bird gets its name from its black-bordered erectile crest, which is orange in the male and yellow in the female. Small groups of Goldcrests can often be seen flying from tree to tree in coniferous forest, drawing attention to themselves by their soft but extremely shrill, high-pitched *zee* calls. They feed mainly on small insects, which they obtain by diligent searching of the foliage or by hovering at the ends of branches, but they have also been seen sipping tree sap. Goldcrests suffer badly in cold weather but after a succession of mild winters they may increase their numbers tenfold. They may lay up to 13 eggs, incubating them for the unusually long period (for passerines) of 14–17 days.

LONG-TAILED TAILORBIRD
Orthotomus sutorius

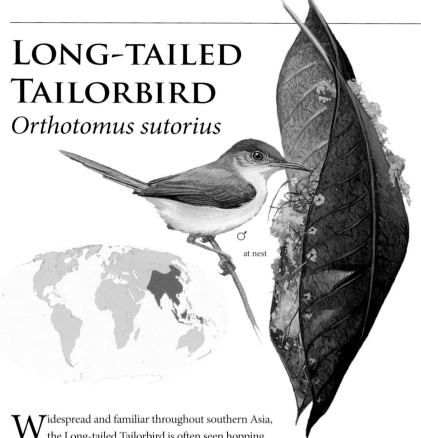

♂

at nest

Widespread and familiar throughout southern Asia, the Long-tailed Tailorbird is often seen hopping about in bushes and around verandas in search of insects and spiders. It has a habit of carrying its tail high over its back and jerking it up and down. The tail is usually longest in the breeding male. Its name is derived from its astonishing nest-building technique. Taking 1 or 2 large leaves on a low bush or branch, it uses its bill as a needle to perforate the edges and sew them together using individual stitches of cottony plant material or the silk from spiders' webs or insect cocoons. The nest itself—made of soft plant fibers—is formed inside this pocket. The 2–3 eggs are incubated by both of the parents.

FACT FILE

RANGE
Indian sub-continent,
Southeast Asia to Java,
S China; up to 5,250 ft
(1,600 m) in Southeast Asia

HABITAT
Thickets, scrub,
bamboo, gardens

SIZE
5 in (12 cm);
breeding male
6 in (15.5 cm)

COMMON TODY FLYCATCHER
Todirostrum cinereum

race *wetmorei*

FACT FILE

RANGE
S Mexico to NW Peru,
Bolivia, SE Brazil

HABITAT
Bushy areas, secondary
growth, thickets, gardens,
overgrown clearings

SIZE
3½–4 in (9–10 cm)

This tiny flycatcher is a common bird, regularly seen in pairs. It forages by moving quickly among leaves to snatch insects, cocking its tail slightly and jerking it from side to side as it flutters and hops within the foliage. Its foraging actions are often more reminiscent of those of wrens than of other tyrant flycatchers. Sometimes it also sallies out after prey. The calls include low, but sharp, chipping notes and a musical trill, which members of a pair use to keep in contact with each other.

Rufous-tailed Plantcutter
Phytotoma rara

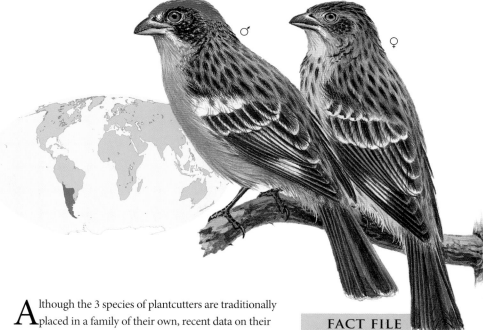

♂ ♀

A lthough the 3 species of plantcutters are traditionally placed in a family of their own, recent data on their biochemistry and natural history suggest that they might be better incorporated in the cotinga family. As their name suggests, they use their thick, serrated bills for cutting or chewing off pieces of leaves, buds, and fruit. Because of this behavior, the birds are sometimes regarded as pests. This species is also known as the Chilean Plantcutter. During the breeding season, these birds remain in pairs, but at other times, they may occur in groups of up to 6 individuals. The males sing with a rasping, metallic wheeze.

FACT FILE

RANGE
C and SC Chile,
extreme W Argentina

HABITAT
Open country with scattered
bushes and thorn trees,
wooded rivers, gardens,
orchards

SIZE
19–20 cm (7½–8 in)

RED-WHISKERED BULBUL

Pycnonotus jocosus

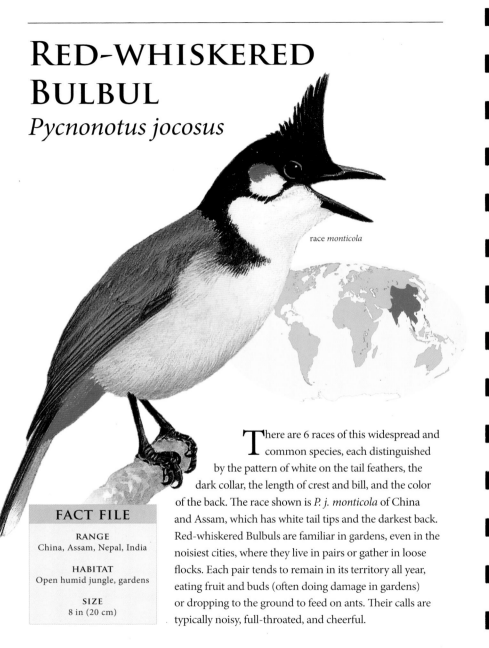

race *monticola*

There are 6 races of this widespread and common species, each distinguished by the pattern of white on the tail feathers, the dark collar, the length of crest and bill, and the color of the back. The race shown is *P. j. monticola* of China and Assam, which has white tail tips and the darkest back. Red-whiskered Bulbuls are familiar in gardens, even in the noisiest cities, where they live in pairs or gather in loose flocks. Each pair tends to remain in its territory all year, eating fruit and buds (often doing damage in gardens) or dropping to the ground to feed on ants. Their calls are typically noisy, full-throated, and cheerful.

FACT FILE

RANGE
China, Assam, Nepal, India

HABITAT
Open humid jungle, gardens

SIZE
8 in (20 cm)

WHITE-CHEEKED BULBUL

Pycnonotus leucogenys

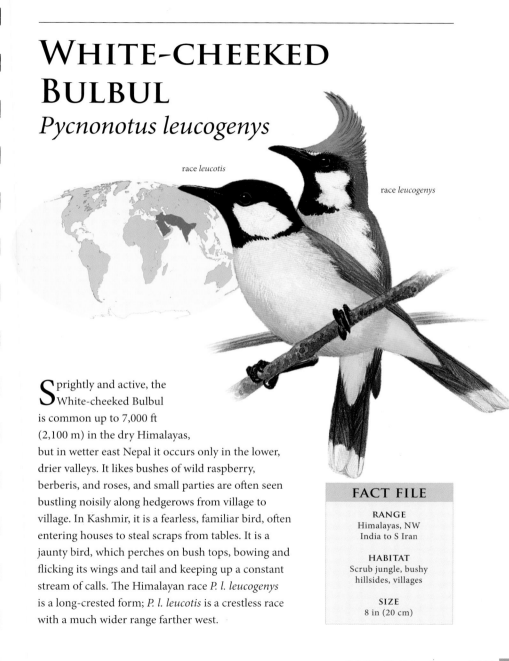

race *leucotis*

race *leucogenys*

Sprightly and active, the White-cheeked Bulbul is common up to 7,000 ft (2,100 m) in the dry Himalayas, but in wetter east Nepal it occurs only in the lower, drier valleys. It likes bushes of wild raspberry, berberis, and roses, and small parties are often seen bustling noisily along hedgerows from village to village. In Kashmir, it is a fearless, familiar bird, often entering houses to steal scraps from tables. It is a jaunty bird, which perches on bush tops, bowing and flicking its wings and tail and keeping up a constant stream of calls. The Himalayan race *P. l. leucogenys* is a long-crested form; *P. l. leucotis* is a crestless race with a much wider range farther west.

FACT FILE

RANGE
Himalayas, NW
India to S Iran

HABITAT
Scrub jungle, bushy
hillsides, villages

SIZE
8 in (20 cm)

Garden Bulbul

Pycnonotus barbatus

race *tricolor*

race *barbatus*

FACT FILE

RANGE
N Africa S to Tanzania

HABITAT
Open forest, gardens,
town parks

SIZE
7 in (18 cm)

All over its large range, the Garden, or Common, Bulbul announces its presence with the high, rich, fluty, or bubbling calls typical of its family. It is an adaptable species, which will thrive almost anywhere where there are buds and fruit, although it does not like dense forest. In some areas, such as Egypt, it has extended its range to take advantage of the increase in irrigated orchards. It is often seen near houses and other buildings and can be very tame, but in more remote areas can be very inconspicuous. The northern race *P. b. barbatus* is white under the tail but the southern race *P. b. tricolor* has a splash of yellow in the same place.

CHESTNUT-EARED BULBUL

Hypsipetes amaurotis

Sociable and noisy, the Chestnut-eared Bulbul visits bird feeders in Japanese gardens and is a familiar sight in parks and wooded areas in towns. The birds that breed in the north move south in winter, often roaming in sizable flocks that communicate with continual loud, fluty, rhythmic calls. In summer, this species is very common in forests on lower mountain slopes. For a bulbul, it is a slim, long-tailed and slender-billed bird, with none of the stubby appearance of many of its African relatives. The race *H. a. amaurotis* is the Japanese form; *H. a. nagamichii* is the race found on Taiwan.

FACT FILE

RANGE
Japan, Taiwan, S China

HABITAT
Forested slopes, gardens, parks

SIZE
10½ in (27 cm)

SPOTTED FLYCATCHER
Muscicapa striata

This rather undistinguished mouse-gray bird has a habit of sitting on a bare branch or other exposed perch, from which it makes short aerial sallies after insects, often returning to the same spot. Its flying style is erratic with many swerves and twists. It flicks its wings or tail when at rest. The Spotted Flycatcher feeds almost entirely on insects but may take berries in the fall. The nest is placed against a wall in a creeper or shrub, on a beam or in a hole. It is a slight structure of moss, wool, or hair, held together with spiders' webs. The clutch is usually 4–5 eggs (rarely 2–7). The eggs are pale blue or green with some brown spots, often around the broad end. Incubation, largely by the female, lasts 12–14 days and fledging takes 11–15 days.

FACT FILE

RANGE
Eurasia N to N Russia and W Siberia, E to N Mongolia, S to NW Africa and Himalayas; winters C and S Africa, Arabia, NW India

HABITAT
Open woods and scrub, parks, gardens; winters in acacia woods and thorn bush

SIZE
5½ in (14 cm)

White-browed Scrubwren
Sericornis frontalis

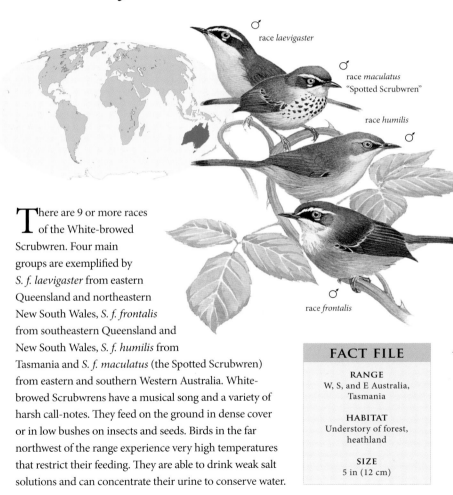

♂
race *laevigaster*

♂
race *maculatus*
"Spotted Scrubwren"

race *humilis*
♂

♂
race *frontalis*

There are 9 or more races of the White-browed Scrubwren. Four main groups are exemplified by *S. f. laevigaster* from eastern Queensland and northeastern New South Wales, *S. f. frontalis* from southeastern Queensland and New South Wales, *S. f. humilis* from Tasmania and *S. f. maculatus* (the Spotted Scrubwren) from eastern and southern Western Australia. White-browed Scrubwrens have a musical song and a variety of harsh call-notes. They feed on the ground in dense cover or in low bushes on insects and seeds. Birds in the far northwest of the range experience very high temperatures that restrict their feeding. They are able to drink weak salt solutions and can concentrate their urine to conserve water.

FACT FILE
RANGE W, S, and E Australia, Tasmania
HABITAT Understory of forest, heathland
SIZE 5 in (12 cm)

WEEBILL
Smicrornis brevirostris

FACT FILE

RANGE
Almost throughout Australia

HABITAT
Eucalypt woodland and
forest; acacia woodland

SIZE
3–3½ in (8–9 cm)

The Weebill is the smallest Australian bird typified by a pale iris and eyebrow and a very short, stubby beak. The song is quite loud for such a tiny bird and resembles *I'm a weebill*. Weebills glean insects from the outer foliage of eucalypts and other trees, often hovering on the edge of the canopy. They eat large numbers of scale insects. They have a long breeding season, from July to February. The nest is round or pear-shaped, with a side entrance and sometimes a tail.

MISTLETOEBIRD
Dicaeum hirundinaceum

This bird has evolved an extremely close association with the 60 or so species of fruiting mistletoe in Australasia. Flowerpeckers generally have a reduced stomach, offset from the gut, but in the Mistletoe bird the stomach has almost completely degenerated. The mistletoe berries pass very rapidly through the intestine where most of the sweet flesh is stripped from the husk, but the seeds are left intact. Seeds may be defecated just 30 minutes after the berries are swallowed. What remains of the bird's stomach is reserved for insects, food that requires a longer period of digestion. Nestlings are fed almost entirely on insects for the first week before being weaned on to the staple diet of berries.

FACT FILE

RANGE
Australia, Aru Islands

HABITAT
Varied, from dry
scrub to wet forest

SIZE
4 in (10 cm)

STRIATED PARDALOTE
Pardalotus striatus

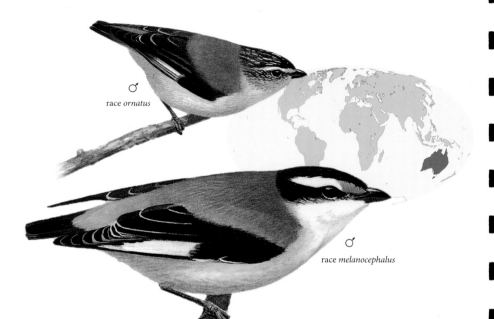

race *ornatus*

♂

race *melanocephalus*

♂

FACT FILE

RANGE
Australia

HABITAT
Forest, woodland,
parks with
eucalypts

SIZE
4 in (10 cm)

The pardolates are small birds that comb the foliage for insects and other food. They feed heavily on "lerps"—sugary secretions exuded by the sap-sucking larvae of psyllid scale insects—and undertake nomadic movements in search of infested trees. The Striated Pardalote is the most widespread species, with several races. There are 2 main groups; black-crowned races, such as the cinnamon-rumped *P. s. melanocephalus* from the northeast, and stripe-crowned races, such as the southeastern race *P. s. ornatus*.

GRAY-BACKED WHITE-EYE
Zosterops lateralis

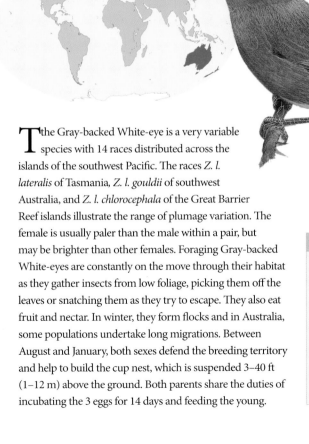

race *lateralis*

The Gray-backed White-eye is a very variable species with 14 races distributed across the islands of the southwest Pacific. The races *Z. l. lateralis* of Tasmania, *Z. l. gouldii* of southwest Australia, and *Z. l. chlorocephala* of the Great Barrier Reef islands illustrate the range of plumage variation. The female is usually paler than the male within a pair, but may be brighter than other females. Foraging Gray-backed White-eyes are constantly on the move through their habitat as they gather insects from low foliage, picking them off the leaves or snatching them as they try to escape. They also eat fruit and nectar. In winter, they form flocks and in Australia, some populations undertake long migrations. Between August and January, both sexes defend the breeding territory and help to build the cup nest, which is suspended 3–40 ft (1–12 m) above the ground. Both parents share the duties of incubating the 3 eggs for 14 days and feeding the young.

FACT FILE

RANGE
Australia, Tasmania, SW
Pacific islands

HABITAT
Practically every type of
vegetation, from rain forest,
mangroves, woods, and
heaths to urban parks and
gardens

SIZE
4–5 in (11–13 cm)

BLUE-FACED HONEYEATER
Entomyzon cyanotis

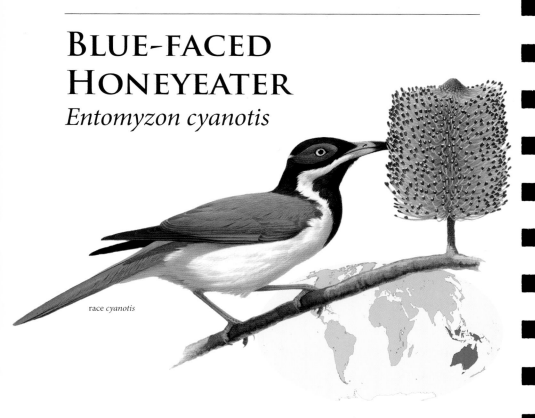

race *cyanotis*

FACT FILE

RANGE
S New Guinea,
N and E Australia

HABITAT
Eucalypt and paperbark
woodland, plantations,
parks, gardens;
occasionally mangroves

SIZE
9½–12 in (24–30 cm)

Well-known for its pugnacious, inquisitive character, the Blue-faced Honeyeater has also earned a bad reputation as a pest of banana plantations, where it feeds on the ripe fruit. It usually forages in small groups, searching for fruit, nectar, and insects in the upper foliage and branches. Groups of 2–10 adults breed together, sharing the brooding and feeding of the 1–2 young. The nest is usually an untidy deep bowl of bark, but occasionally the birds will take over deserted nests of other birds. The northwestern race *E. c. albipennis* has white wing patches, which are conspicuous in flight. The eastern race *E. c. cyanotis* has only a small buff patch on its wings.

NOISY FRIARBIRD
Philemon corniculatus

This large honeyeater is also known as the Knobby-nosed Leatherhead because of the black naked skin on its head and the small knob at the base of its upper mandible. It often gathers in groups at flowering trees, squabbling raucously over the spoils as it feeds on fruit and collects the nectar of banksia and eucalypt trees. It also consumes insects, which provide valuable protein in an otherwise sugar-rich diet. In the breeding season (July to January), the flocks break up, and the birds advertise their feeding and breeding territories with grating, discordant calls. The females incubate the 2–3 eggs alone, but both sexes feed the young.

FACT FILE

RANGE
S New Guinea, E Australia

HABITAT
Open forest, woodland, and lowland savanna

SIZE
12–13 in (30–34 cm)

YELLOW WATTLEBIRD
Anthochaera paradoxa

imm

Named for the bright yellow wattles that hang from its cheeks, the Yellow Wattlebird is the largest of the honeyeaters and was once shot as a gamebird. Outside of the breeding season, it often lives in flocks of 10–12 birds; these may be sedentary or nomadic, foraging through the canopy for insects, gleaning them from the leaves, branches, and bark. In addition, the birds eat nectar and fruit. Yellow Wattlebirds breed as pairs, establishing territories in August and laying 2–3 eggs in an untidy nest built of thin twigs and grasses, placed in thick foliage.

CARDINAL HONEYEATER
Myzomela cardinalis

In most honeyeaters, the sexes are similar but the Cardinal Honeyeater is an exception. Whereas the female is dull olive-gray with small red patches, the male has bright scarlet and black plumage. His flamboyant appearance is matched by his noisy, aggressive territorial behavior. These dramatic-looking birds normally feed in the canopy but they venture lower down when near the forest edge, sipping nectar from flowers and gleaning insects from the foliage. They may breed at any time of year, suspending a fragile nest of plant fiber from a tree fork and laying 2–3 eggs, which are incubated by both parents.

FACT FILE

RANGE
Islands of Vanuatu, Samoa, Santa Cruz, and Solomons

HABITAT
Coastal scrub, secondary forest, woodland, gardens

SIZE
15 in (13 cm)

New Holland Honeyeater

Phylidonyris novaehollandiae

Tew Holland or Yellow-winged,
Honeyeaters are aggressive
birds but their aggression varies
along with the abundance of nectar
and nesting resources. They catch
insects for their protein content but
rely on honeydew and nectar to supply
energy. As they gather the nectar, they are
dusted with pollen, which is transferred to
successive blossoms and in this way, they play
an important role in the pollination of a wide
range of Australian plants. They may breed up to 3
times a year, depending on the availability of nectar. The
males defend nest sites that are arranged in loose colonies,
usually 3–6 ft (1–2 m) above the ground in shrubs. The
females build the nests and incubate the 2–3 eggs, but the
young are fed by both parents.

FACT FILE

RANGE
SW and E Australia,
Tasmania

HABITAT
Heathland, mallee heath,
woods, and forest with
shrubs

SIZE
6½–7½ in (17–19 cm)

WESTERN SPINEBILL
Acanthorhynchus superciliosus

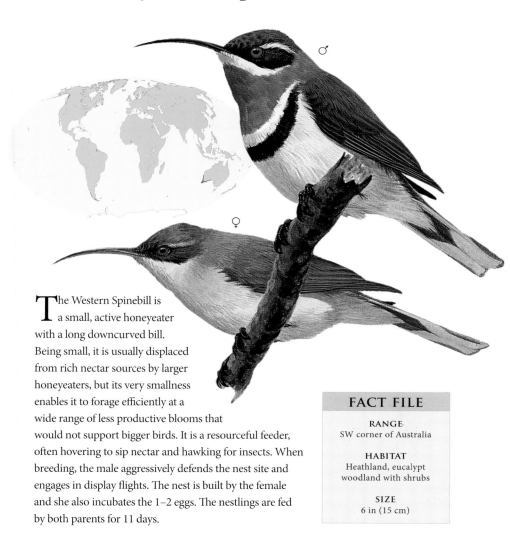

The Western Spinebill is a small, active honeyeater with a long downcurved bill. Being small, it is usually displaced from rich nectar sources by larger honeyeaters, but its very smallness enables it to forage efficiently at a wide range of less productive blooms that would not support bigger birds. It is a resourceful feeder, often hovering to sip nectar and hawking for insects. When breeding, the male aggressively defends the nest site and engages in display flights. The nest is built by the female and she also incubates the 1–2 eggs. The nestlings are fed by both parents for 11 days.

FACT FILE

RANGE
SW corner of Australia

HABITAT
Heathland, eucalypt woodland with shrubs

SIZE
6 in (15 cm)

Noisy Miner
Manorina melanocephala

FACT FILE

RANGE
E Australia, Tasmania

HABITAT
Open forest, woodland, and
partly cleared land

SIZE
10–11½ in (25–29 cm)

Noisy Miners have a complex social system with territorial groups of 6–30 birds gathering together into large colonies. The group members cooperate to mob predators and they will aggressively exclude most other bird species from the nesting area. They breed communally, with as many as 12 males caring for the young of each polyandrous female. The nest, which is built by the female, is a loosely constructed bowl of twigs and grasses placed in the fork of a shrub or tree.

AFRICAN YELLOW WHITE-EYE

Zosterops senegalensis

This is the typical, common white-eye of Africa. Widespread and frequently encountered in open woodland or among scattered trees, it is rarely seen in dense forest. Small parties often mix with sunbirds and other species, flitting through the canopy and down into lower bushes in a constant search for the insects that make up most of their diet. The birds maintain contact with one another by means of weak, twittering calls and piping notes as they fly from tree to tree. The pair builds a nest of fine plant fiber decorated with lichens, either placed in a tree fork or suspended from a branch.

FACT FILE

RANGE
Sub- Saharan Africa

HABITAT
Open thornbush,
dry woods, gardens

SIZE
4 in (10 cm)

PYGMY SUNBIRD
Anthreptes platurus

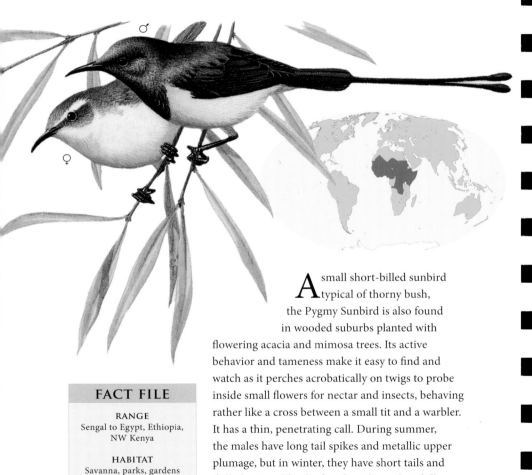

FACT FILE

RANGE
Sengal to Egypt, Ethiopia,
NW Kenya

HABITAT
Savanna, parks, gardens

SIZE
6–7 in (16–18 cm)

A small short-billed sunbird typical of thorny bush, the Pygmy Sunbird is also found in wooded suburbs planted with flowering acacia and mimosa trees. Its active behavior and tameness make it easy to find and watch as it perches acrobatically on twigs to probe inside small flowers for nectar and insects, behaving rather like a cross between a small tit and a warbler. It has a thin, penetrating call. During summer, the males have long tail spikes and metallic upper plumage, but in winter, they have short tails and dull plumage, like the females. This rather dull, non-breeding eclipse plumage is a characteristic of many sunbirds.

SCARLET-CHESTED SUNBIRD

Nectarinia senegalensis

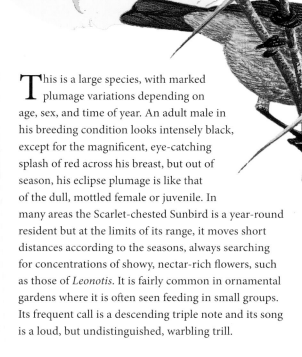

This is a large species, with marked plumage variations depending on age, sex, and time of year. An adult male in his breeding condition looks intensely black, except for the magnificent, eye-catching splash of red across his breast, but out of season, his eclipse plumage is like that of the dull, mottled female or juvenile. In many areas the Scarlet-chested Sunbird is a year-round resident but at the limits of its range, it moves short distances according to the seasons, always searching for concentrations of showy, nectar-rich flowers, such as those of *Leonotis*. It is fairly common in ornamental gardens where it is often seen feeding in small groups. Its frequent call is a descending triple note and its song is a loud, but undistinguished, warbling trill.

FACT FILE

RANGE
W and E Africa from Senegal to Kenya and S to Zimbabwe

HABITAT
Woodland, parks, riverside scrub

SIZE
6 in (15 cm)

SUPERB SUNBIRD
Nectarinia superba

♂

♀

In poor light, the male Superb Sunbird looks almost black, but in full sunshine his metallic plumage dazzles the eye with its breathtakingly beautiful coloration. These birds are usually found in sunny places at the edges of forests, although they also visit savanna woods and thick bush as long as there is plenty of nectar-rich blossom for them to feed from. These stocky, fairly large sunbirds are hard to follow as they fly swiftly from tree to tree, bounding and zigzagging. Although the species has an extensive range, it is common nowhere, and the birds' chirping calls and poor, jingling song are of little help in locating them.

FACT FILE

RANGE
Sierra Leone E to Central
African Republic, S to Congo

HABITAT
Forest edge and clearings

SIZE
6 in (15 cm)

Yellow-bellied Sunbird

Nectarinia jugularis

♂

♀

The Yellow-bellied, or Olive-backed, Sunbird is the only sunbird found in Australia, where it is a common, locally nomadic species in the forests and gardens of northeast Queensland. It is an active bird, which searches a large variety of native and exotic plants for nectar and insects. It usually clings to foliage while feeding but may hover briefly to sip nectar from a flower or pluck a spider from its web. Like most sunbirds, it is incapable of the sustained hovering practiced by hummingbirds, which are the ecological equivalents of sunbirds in the New World. Males and females are fairly similar but the female's underparts are yellow throughout whereas the male has a metallic purplish-blue throat and chest. The male defends the breeding territory but the female builds the nest: this has a "tail" of loose material.

FACT FILE

RANGE
Southeast Asia, from Indonesia to Solomon Islands, NE Australia

HABITAT
Coastal rain forest, mangroves, and gardens

SIZE
4–5 in (10–12 cm)

NORTHERN ORANGE-TUFTED SUNBIRD

Nectarinia osea

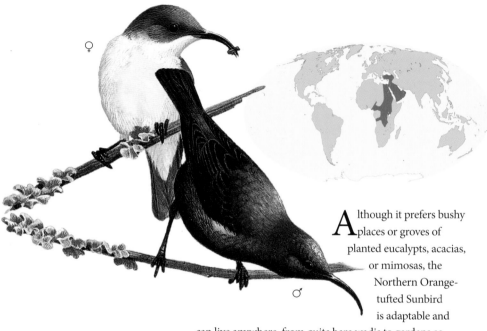

A lthough it prefers bushy places or groves of planted eucalypts, acacias, or mimosas, the Northern Orange-tufted Sunbird is adaptable and can live anywhere, from quite bare wadis to gardens as long as they have tiny insects or nectar-bearing flowers. It is resident throughout its range except for some winter dispersal. Its song is a shrill, rapid, rambling trill, quite unlike that of any other bird in its range. From July to November, the mature males look like dull females, but they often have a few traces of their breeding colors on the wings or throat. This is common among sunbirds in eclipse plumage. The orange tufts flanking the breast are visible only in display.

FACT FILE

RANGE
Syria, Israel, W and S edge of Arabian peninsula, parts of sub-Saharan Africa

HABITAT
Woods, orchards, gardens

SIZE
4 in (11 cm)

SCARLET-BACKED FLOWERPECKER
Dicaeum cruentatum

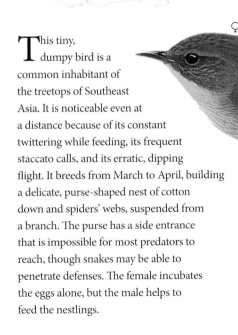

This tiny, dumpy bird is a common inhabitant of the treetops of Southeast Asia. It is noticeable even at a distance because of its constant twittering while feeding, its frequent staccato calls, and its erratic, dipping flight. It breeds from March to April, building a delicate, purse-shaped nest of cotton down and spiders' webs, suspended from a branch. The purse has a side entrance that is impossible for most predators to reach, though snakes may be able to penetrate defenses. The female incubates the eggs alone, but the male helps to feed the nestlings.

FACT FILE

RANGE
India to S China, Southeast
Asia, Indonesia

HABITAT
Coastal lowland plains, open
forest, scrub, gardens

SIZE
4 in (10 cm)

PURPLE-RUMPED SUNBIRD

Nectarinia zeylonica

♂

♀

FACT FILE

RANGE
Peninsular India,
Bangladesh, Sri Lanka

HABITAT
Jungle and dry cultivated
plains with deciduous trees

SIZE
4 in (10 cm)

Several sunbirds have evolved long bills, which enable them to drink nectar from long, tubular flowers. Purple-rumped Sunbirds, instead, use their short bills to pierce large flowers neatly at the base of the petals to reach the nectar. Males and females differ. Though young males look like rather yellow females, the adult males keep their metallic plumage all year round. They are usually found in pairs, actively flitting among thick foliage and flowers, perching acrobatically to reach into the flowers or—more rarely—hovering briefly to take a fly. The pair will defend their feeding tree and nesting area vigorously if other sunbirds appear. The song is a sharp twitter or a squeaky performance given with flicking wings and tail.

Yellow-backed Sunbird
Aethopyga siparaja

The Yellow-backed, or Crimson, Sunbird is a common resident throughout its range. In the Nepal region, it moves higher up the mountains in summer and descends again in winter, when it often visits gardens. Unlike many sunbirds, the male Yellow-backed Sunbird retains his bright plumage throughout the year. It often forages alone, taking nectar from a wide variety of trees and ornamental flowers and adapting its feeding technique to suit the flower. It will pierce larger blooms, such as hibiscus, to reach the nectar through the base of the petals, but will drink from tubular flowers by hovering before them, rather like a hummingbird, and inserting its tubular tongue. It seems to prefer bright red flowers to any others.

FACT FILE

RANGE
Sumatra, Borneo, Malaysia,
W to India

HABITAT
Gardens, orchards,
pine forest

SIZE
4 in (10 cm)

BLEATING CAMAROPTERA
Camaroptera brachyura

race *brachyura*
"Green-backed
Camaroptera"

race *brevicaudata*
"Gray-backed
Camaroptera"

FACT FILE

RANGE
Sub-Saharan Africa

HABITAT
Woodland thickets,
forest edge, riverine bush,
parks, gardens

SIZE
5 in (12.5 cm)

Of the half dozen or so camaropteras found in Africa, this is the most common. A green-backed group of races, including *C. b. brachyura*, occurs mostly down the eastern edge of Africa, from Kenya south to South Africa. A gray-backed group, *C. b. brevicaudata*, is widespread elsewhere. This is often regarded as a separate species, the Gray-backed Camaroptera *C. brevicaudata*. The green-backed birds prefer moist evergreen forests, whereas the gray-backed ones favor dry thornveld and open broad-leaved woodlands. They are not easy to see as they forage among the vegetation but their alarm call is quite distinctive, like the bleating of a lamb. Another call sounds like tapping stones. Its soft, downy nest is made within a frame formed from the broad leaves of a single twig or spray, with more leaves added to make a roof.

NORTHERN CROMBEC
Sylvietta brachyura

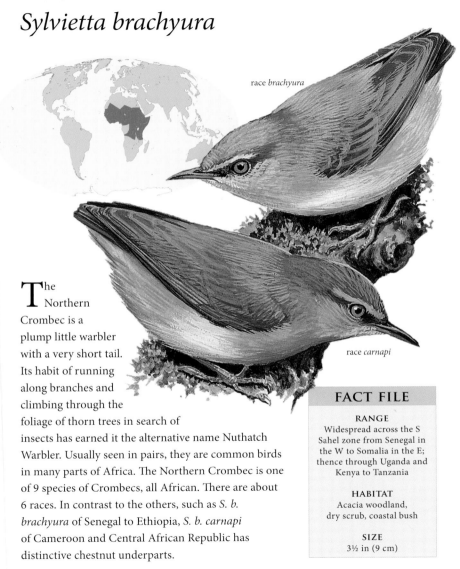

race *brachyura*

race *carnapi*

The Northern Crombec is a plump little warbler with a very short tail. Its habit of running along branches and climbing through the foliage of thorn trees in search of insects has earned it the alternative name Nuthatch Warbler. Usually seen in pairs, they are common birds in many parts of Africa. The Northern Crombec is one of 9 species of Crombecs, all African. There are about 6 races. In contrast to the others, such as *S. b. brachyura* of Senegal to Ethiopia, *S. b. carnapi* of Cameroon and Central African Republic has distinctive chestnut underparts.

FACT FILE

RANGE
Widespread across the S Sahel zone from Senegal in the W to Somalia in the E; thence through Uganda and Kenya to Tanzania

HABITAT
Acacia woodland, dry scrub, coastal bush

SIZE
3½ in (9 cm)

ICTERINE WARBLER
Hippolais icterina

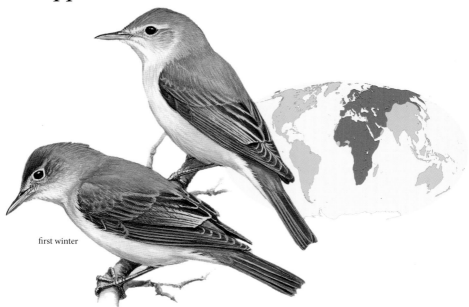

first winter

FACT FILE

RANGE
N and E Europe S to Alps, Asia Minor, and Caucasus, Asia E to Altai Mountains; winters E and tropical Africa

HABITAT
Open broad-leaved woods, parks, riversides, farmland, orchards, and large gardens

SIZE
5 in (13 cm)

The tree warblers of the genus *Hippolais* are relatively large, heavily built birds with prominent legs and feet and a habit of flicking their tails. The Icterine Warbler is one of the more striking members of the group, with bright yellow underparts and a powerful mimetic song. In the fall, its plumage is duller. It is a lively, active warbler, more conspicuous than many as it moves about the tree canopy in search of insects and their larvae. In the fall, it will also take fruit, especially berries. The nest of the Icterine Warbler is a fine example of tree architecture. Built in the fork of a tree, it is made of stems or grasses held together with wool and spiders' silk. Some nests incorporate bark, paper, or even rags. The 4–5 eggs are variable in size.

BLACKCAP
Sylvia atricapilla

race *atricapilla*

T his active, lively bird is often easy to observe in the trees as it combs the foliage for insects. The glossy black cap of the male and the reddish-brown cap of the female are distinctive. The male has a wavy or rippling song full of rich, pure notes; he is also an accomplished mimic. Before the females arrive on the breeding grounds, the male often builds several rudimentary "cock's nests" of dry grass or roots, each slung between 2 twigs, up to 10 ft (3 m) from the ground. The female may reject them all and build one of her own— sometimes with the male's help—or she may accept one of his, finishing it off with a lining of down or wool. Both adults incubate the 4–6 eggs, which vary greatly in color and markings. The race *S. a. heineken* found on Madeira and the Canary Islands is darker than the European race *S. a. atricapilla*.

FACT FILE

RANGE
Eurasia E to River Irtysh in Siberia and N Iran, NW Africa and Atlantic islands; winters Mediterranean and Africa S to Tanzania

HABITAT
Broad-leaved and conifer woods, overgrown hedges, scrub with tall trees

SIZE
5½ in (14 cm)

WILLOW WARBLER
Phylloscopus trochilus

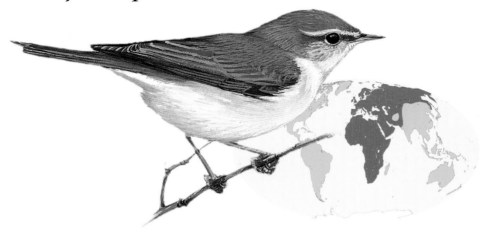

The Willow Warbler is one of about 45 species of leaf warblers in the genus *Phylloscopus*—they are all small, slender, graceful birds with basically green, yellow, or brown plumage. The Willow Warbler is almost identical to the Chiffchaff *P. collybita* in appearance but its song is completely different with a descending series of liquid, silvery notes, in contrast to the chiffchaff's 2-note repetition of its name. It is an adaptable bird, which will thrive in any type of open woodland and its song is a common sound in many regions. It feeds on insects taken on the wing or picked off the foliage with delicate precision. In winter, when the insect supply fails in the north, it flies south to Africa. The nest is a domed structure with a side entrance, built on or near the ground by the female. If disturbed on the nest, the female may feign injury, fluttering along the ground to lure the intruder away from the eggs or nestlings.

FACT FILE

RANGE
Scandinavia and NW
Europe to E Siberia and
Alaska; winters tropical
and S Africa

HABITAT
Open woods, scrub,
conifer, plantations,
moorland with
shrubs, hedges

SIZE
4 in (10.5 cm)

JAPANESE WHITE-EYE
Zosterops japonica

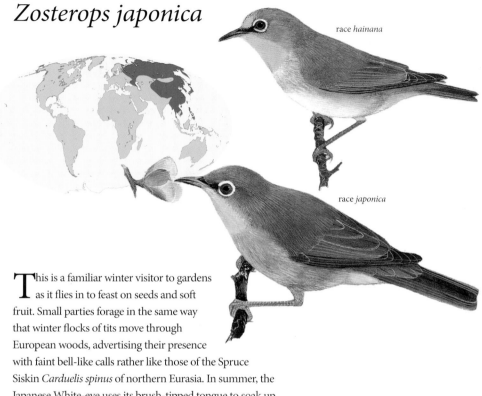

race *hainana*

race *japonica*

This is a familiar winter visitor to gardens as it flies in to feast on seeds and soft fruit. Small parties forage in the same way that winter flocks of tits move through European woods, advertising their presence with faint bell-like calls rather like those of the Spruce Siskin *Carduelis spinus* of northern Eurasia. In summer, the Japanese White-eye uses its brush-tipped tongue to soak up nectar from large blooms, but its main diet consists of insects and spiders, which it gleans from tree foliage and picks out of bark crevices with its short, fine-pointed bill. Although these birds are regarded as pests by commercial fruit-growers, any damage they do in orchards is far outweighed by the number of insect pests they eat. The race *Z. j. japonica* is found on Honshu Island, Japan. The distinctive *Z. j. hainana*, found on Hainan Island off southern China, is small, with brighter, more yellowish plumage.

FACT FILE

RANGE
China, Indochina, Japan

HABITAT
Low hill forest, parks, and gardens

SIZE
4 in (11 cm)

ORIENTAL WHITE-EYE
Zosterops palpebrosa

race *gouldii*

Typically small, dumpy white-eyes, with a good deal of yellow, white, and green in their plumage, Oriental White-eyes live in small groups, which spend much of their time in the trees, searching for weevils, ants, and other insects, as well as their eggs. The parties set up a high, querulous chorus of calls as they feed, repeating the song over and over again with no variation. In forests and cultivated areas, they frequently forage in mixed-species flocks with other birds. The male and female build the nest together, weaving a delicate cup of grass and spiders' webs suspended in the fork of a tree or shrub. The pale blue eggs are incubated for 12 days and the young leave the nest 12 days after hatching, freeing the parents to produce a second brood.

FACT FILE

RANGE
Afghanistan E to China,
Malaysia, Indonesia

HABITAT
Hill forest,
mangroves, gardens

SIZE
4 in (10 cm)

YELLOW-RUMPED WARBLER

Dendroica coronata

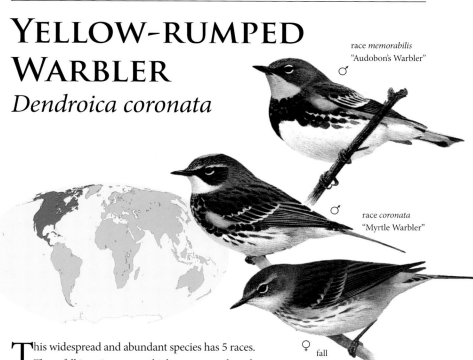

race *memorabilis*
"Audobon's Warbler"

♂

♂

race *coronata*
"Myrtle Warbler"

♀ fall

This widespread and abundant species has 5 races. These fall into 2 groups, which were once thought to be separate species; the white-throated Myrtle Warbler, typified by *D. c. coronata* of Canada and central and eastern U.S.A. and the yellow-throated Audubon's Warbler, typified by *D. c. auduboni* of southwest Canada and western U.S.A. The race illustrated, *D. c. memorabilis*, from southwest U.S.A., is a particularly strongly marked form of the Audubon Warbler group. Apart from their plumage differences, most noticeable in breeding males, their call notes are also distinct, that of Audubon's Warbler being softer than that of its eastern counterpart. Both nest in conifers and tall deciduous trees. They may interbreed where their ranges overlap. In summer, both races feed largely on insects, but in winter and during migration, they take a much wider range of foods.

FACT FILE

RANGE
Breeds North America,
S to C and S Mexico and
Guatemala; winters S of
breeding range,
to Central America

HABITAT
Coniferous and
deciduous woods

SIZE
5–6 in (13–15 cm)

AMERICAN REDSTART
Setophaga ruticilla

FACT FILE

RANGE
Breeds SE Alaska, E through
C Canada, S through Texas
to E U.S.A.; winters extreme
S U.S.A. to Brazil

HABITAT
Deciduous and
mixed woodland

SIZE
4–5 in (11–13.5 cm)

Few warblers are as vividly patterned or as bold or animated as the New World redstarts. In most of the 11 other species, placed in a different genus, *Myioborus*, the sexes are alike, but the female American Redstart is quite unlike her mate, with incandescent yellow where he has flaming orange. The dazzling colors illuminate every flutter of wings and tail as the birds work through the foliage in search of caterpillars, or dart into the air like flycatchers to catch insects on the wing. Their songs feature shrill, sibilant, slurred notes and they often utter sharp, staccato calls. The female builds the cup nest alone and places it in the fork of a tree or shrub, often decorating the lining with the gaudy feathers of brightly colored buntings or tanagers.

WHITE-BROWED FANTAIL

Rhipidura aureola

race *aureola*

There are 3 races of this flycatcher: *R. a. burmanica* in the east, with its crown and back of similar color, *R. a. aureola* across the bulk of India, with a darker cap, and *R. a. compressirostris* in the south and Sri Lanka, with less white in the tail. Sexes are similar, but females are generally a little paler on the head than the males. Restless, alert birds, White-browed Fantails are usually seen in ones or twos but sometimes join mixed parties of birds roving through low bushes and undergrowth. The tail is often fanned and held up high, with the wings drooped. The species is thought to use its flicking wings and tail to disturb insects from crevices in old bark. The insects are caught in midair in graceful, swooping flights from a perch. The song is a thin, clear whistle, rising and descending in a short phrase. A compact nest of plant fiber and grass, covered with spiders' webs, is placed on a high branch; 3 eggs are laid.

FACT FILE

RANGE
Sri Lanka, India, Pakistan, Bangladesh

HABITAT
Forests, gardens, scrub

SIZE
6½ in (17 cm)

TOP 100 U.S. BACKYARD BIRDS

Acorn Woodpecker
Size: 7½–9 in
Diet: Insects, acorns, sap, fruit
Range: W and S U.S.A., Central America

American Coot
Size: 16 in
Diet: Plants
Range: S Canada, U.S.A., Central America, Caribbean

American Crow
Size: 17–18 in
Diet: Insects, seeds, berries, small amphibians and reptiles, mice, eggs and nestlings of other species, carrion
Range: North America

American Goldfinch
Size: 4 in
Diet: Seeds, occasionally insects
Range: S Canada, U.S.A.

American Kestrel
Size: 9–12 in
Diet: Large insects, small mammals, reptiles, amphibians, birds
Range: SE Alaska and Canada to South America

American Robin
Size: 9–11 in
Diet: Worms and other invertebrates, fruit
Range: North America

American Tree Sparrow
Size: 6 in
Diet: Seeds during winter, insects during summer
Range: Alaska, Canada, U.S.A.

Anna's Hummingbird
Size: 4 in
Diet: Nectar, young also eat small insects
Range: W U.S.A.

Baltimore Oriole
Size: 9 in
Diet: Insects, berries, nectar
Range: S and C Canada, E U.S.A., Mexico, N South America

Band-tailed Pigeon
Size: 13–16 in
Diet: Seeds, fruit, acorns, pine nuts, flowers
Range: W and S U.S.A., parts of Central America and South America

Barn Owl
Size: 12½–16 in
Diet: Small mammals
Range: U.S.A., Central America, South America, parts of the Caribbean

Black-billed Magpie
Size: 18 in
Diet: Insects, carrion, eggs and young of other birds, some plant products
Range: North America

Black-capped Chickadee
Size: 5 in
Diet: Insects, seeds
Range: Alaska, Canada, S to C U.S.A.

Black-chinned Hummingbird
Size: 3½ in
Diet: Nectar, small insects and spiders
Range: W U.S.A., Mexico

Black-headed Grosbeak
Size: 8 in
Diet: Insects, seeds, berries
Range: SW Canada, W U.S.A., Mexico

Black Phoebe
Size: 6½ in
Diet: Insects, some small berries and small fish
Range: SW U.S.A., Central America, NW South America

Blue Jay
Size: 12 in
Diet: Nuts, fruit, seeds, insects, small vertebrates
Range: E North America, from S Canada to Mexican Gulf

Bohemian Waxwing
Size: 8 in
Diet: Insects during summer, berries during fall and winter
Range: North America

Brewer's Blackbird
Size: 8½–10 in (male), 8–9 in (female)
Diet: Seeds, grain; insects during summer; occasionally small frogs, young voles, nestling birds
Range: S Canada, U.S.A., Mexico

Brown Thrasher
Size: 10 in
Diet: Insects, berries, small frogs, snakes
Range: E U.S.A., S Canada

Brown-headed Cowbird
Size: 7–8 in
Diet: Seeds, insects
Range: Alaska, W Canada, N U.S.A.

Bushtit
Size: 4 in
Diet: Insects, some spiders and berries
Range: W U.S.A., Mexico

California Towhee
Size: 9 in
Diet: Seeds and insects on the ground, occasionally flying insects
Range: W U.S.A.

Carolina Chickadee
Size: 4–5 in
Diet: Insects, spiders, seeds, berries
Range: SE U.S.A.

Carolina Wren
Size: 6–7 in
Diet: Insects, small spiders, some seeds
Range: E U.S.A.

Cassin's Finch
Size: 6 in
Diet: Seeds, buds, berries
Range: SW Canada, W U.S.A., Mexico

Cedar Waxwing
Size: 6–7 in
Diet: Berries, insects
Range: U.S.A., Mexico

Chestnut-backed Chickadee
Size: 5 in
Diet: Insects, seeds, also berries
Range: NW Canada, S to California

Chimney Swift
Size: 5 in
Diet: Flying insects
Range: S Canada, E U.S.A, Central America, NE South America

Chipping Sparrow
Size: 5–5½ in
Diet: Insects, small seeds, fruit
Range: Canada, S to N Nicaragua

Cliff Swallow
Size: 5 in
Diet: Flying insects
Range: S Alaska, S Canada, U.S.A., Central America, N to C South America, Caribbean

Common Grackle
Size: 11–13 in
Diet: Insects, seeds, acorns, fruit, sometimes small birds, mice, frogs
Range: C to E U.S.A. and Canada

Common Nighthawk
Size: 9½ in
Diet: Insects
Range: North America (except extreme N), South America, West Indies

Common Raven
Size: 24 in
Diet: Insects, rodents, amphibians, eggs, young birds, carrion
Range: W and N North America, Central America

Common Redpoll
Size: 5 in
Diet: Seeds, some insects during summer
Range: Alaska, Canada, N U.S.A.

Cooper's Hawk
Size: 16 in
Diet: Birds, small mammals
Range: S Canada, U.S.A., Mexico

Dark-eyed Junco
Size: 6 in
Diet: Seeds, insects
Range: Canada, N and C U.S.A., S to Mexico

Downy Woodpecker
Size: 6–7 in
Diet: Insects, fruit, seeds
Range: Alaska, Canada, U.S.A.

Eastern Bluebird
Size: 5½–7½ in
Diet: Insects, fruit, berries
Range: E North America, Central America

Eastern Phoebe
Size: 7 in
Diet: Flying insects, occasionally small fruit
Range: N and E North America, SE and SC U.S.A. to Mexico

Eastern Screech Owl
Size: 7½–9 in
Diet: Insects, crayfish, earthworms, songbirds, rodents
Range: E North America to Florida and NE Mexico

Eastern Towhee
Size: 8 in
Diet: Insects, seeds, berries
Range: E U.S.A.

Eastern Wood-pewee
Size: 6 in
Diet: Flying insects
Range: C to E U.S.A., E Mexico, Central America, NW South America, Caribbean

European Starling
Size: 8 in
Diet: Insects, fruit, seeds
Range: North America

Evening Grosbeak
Size: 8 in
Diet: Seeds, some insects, berries
Range: W North America, S to Mexico, E across Canada

Fox Sparrow
Size: 7 in
Diet: Seeds, insects
Range: W Canada, U.S.A.

Golden-crowned Kinglet
Size: 3–4½ in
Diet: Small insects and their eggs
Range: S Alaska, S Canada, U.S.A.

Golden-crowned Sparrow
Size: 7 in
Diet: Insects, seeds
Range: W Canada and U.S.A.

Great Crested Flycatcher
Size: 6½–8½ in
Diet: Insects, other invertebrates, some small fruit
Range: Parts of S Canada, C to E U.S.A., E Mexico, Central America, N South America, Caribbean

Hairy Woodpecker
Size: 7–9½ in
Diet: Insects
Range: Alaska, Canada, S to Central America

Harris's Sparrow
Size: 7 in
Diet: Seeds, insects, berries
Range: N Canada to S U.S.A.

Herring Gull
Size: 22–26 in
Diet: Fish, marine invertebrates, insects, birds, eggs, carrion, garbage
Range: Canada, U.S.A.

House Finch
Size: 6 in
Diet: Seeds, berries
Range: North America, Central America, S to Mexico

House Sparrow
Size: 5½–7 in
Diet: Seeds, some insects
Range: North America

House Wren
Size: 4½–5 in
Diet: Insects and spiders, snail shells
Range: S Canada, U.S.A., Central America, South America

Inca Dove
Size: 8 in
Diet: Seeds
Range: SW U.S.A. to N Costa Rica

Indigo Bunting
Size: 5 in
Diet: Seeds, insects
Range: E and SW U.S.A., Central America

Ladder-backed Woodpecker
Size: 6½–7 in
Diet: Insects and arthropods
Range: SW U.S.A., Mexico

Least Flycatcher
Size: 4½–5½ in
Diet: Insects, some fruit during winter
Range: S Canada, C to E U.S.A., Central America

Lesser Goldfinch
Size: 4 in
Diet: Seeds, some insects
Range: SW and W U.S.A., Central America, N South America

Mallard
Size: 20–25½ in
Diet: Insects, larvae, aquatic invertebrates and vegetation, seeds, acorns, grain
Range: North America

Mountain Chickadee
Size: 5 in
Diet: Insects, seeds, berries
Range: W Canada and U.S.A.

Mourning Dove
Size: 12 in
Diet: Seeds
Range: SE Alaska, S Canada to C Panama, Caribbean

Northern Cardinal
Size: 8½ in
Diet: Seeds, insects, berries
Range: SE Canada, E, C, and SW U.S.A., Mexico to Belize; Hawaii

Northern Flicker
Size: 10–14 in
Diet: Ants and other insects, fruit, berries
Range: North America, S to Nicaragua

Northern Mockingbird
Size: 9–11 in
Diet: Insects, berries
Range: S Canada, U.S.A., Mexico, Caribbean, Bermuda, Hawaii

Oak Titmouse
Size: 6 in
Diet: Insects, seeds
Range: SW U.S.A., Oregon to California

Osprey
Size: 21½–23 in
Diet: Fish
Range: North
America

Peregrine Falcon
Size: 14–19½ in
Diet: Birds, from
songbirds to
small geese;
bats and other
small mammals
Range: Alaska,
N and W Canada,
W and S U.S.A.,
Central America,
South America,
Caribbean

**Pileated
Woodpecker**
Size: 16 in
Diet: Ants and
other insects,
nuts, fruit
Range: Parts of
Canada, W U.S.A.,
E U.S.A.

**Pine
Grosbeak**
Size: 8 in
Diet: Buds,
berries, seeds,
insects
Range: N and
W North America

**Pine
Siskin**
Size: 5 in
Diet: Seeds,
some insects
Range: North
America

**Purple
Finch**
Size: 6 in
Diet: Seeds,
buds, also
berries and
insects
Range: Parts of
Canada, W and
E U.S.A.

**Purple
Martin**
Size: 7 in
Diet: Flying
insects
Range: North
America,
N South
America

**Red-bellied
Woodpecker**
Size: 9 in
Diet: Insects,
nuts, fruit
Range: E U.S.A.

**Red-breasted
Nuthatch**
Size: 4 in
Diet: Insects,
seeds
Range: S Canada,
U.S.A.

**Red-eyed
Vireo**
Size: 5½ in
Diet: Insects
(especially
caterpillars),
small fruit
Range: Canada
to S, NW South
America

**Red-headed
Woodpecker**
Size: 7½ in
Diet: Beech and
oak mast, seeds,
nuts, berries,
fruit, insects,
bird's eggs,
nestlings, mice
Range: S Canada,
E U.S.A.

**Red-winged
Blackbird**
Size: 9 in
Diet: Insects,
seeds
Range: Parts of
Canada, U.S.A.,
Central America

**Ring-billed
Gull**
Size: 17–21½ in
Diet: Fish, insects,
rodents earthworms,
grain, garbage
Range: S Canada,
U.S.A., Mexico,
Caribbean

Rock Pigeon
Size: 12 in
Diet: Seeds, grain
Range: S Canada,
U.S.A.

**Rose-breasted
Grosbeak**
Size: 8 in
Diet: Insects,
seeds, berries
Range: SC Canada,
E U.S.A, S to
Mexico and N
South America

Ruby-crowned Kinglet
Size: 4–4½ in
Diet: Insects, insect eggs, occasionally berries
Range: NW Alaska, E Canada, U.S.A, N Mexico

Ruby-throated Hummingbird
Size: 3½ in
Diet: Nectar, some insects and spiders
Range: E North America, coastal parts of SE U.S.A. to NW Costa Rica

Rufous Hummingbird
Size: 4 in
Diet: Nectar
Range: SE Alaska, W Canada and U.S.A., Mexico

Sharp-shinned Hawk
Size: 10–13 in
Diet: Small birds, also small mammals and snakes
Range: North America, Central America, South America, Caribbean

Song Sparrow
Size: 5½ in
Diet: Insects, seeds
Range: S Alaska, C Canada S to N Mexico

Spotted Towhee
Size: 8 in
Diet: Seeds, insects, berries
Range: W to C U.S.A

Steller's Jay
Size: 11 in
Diet: Seeds, nuts, berries, insects
Range: W Canada and U.S.A., Central America

Tufted Titmouse
Size: 6½ in
Diet: Insects, seeds
Range: E North America

Turkey Vulture
Size: 25–32 in
Diet: Carrion, from small mammals to dead cows, also some insects, other invertebrates, and fruit
Range: U.S.A., Central America, South America, Caribbean

Varied Thrush
Size: 9 in
Diet: Insects, berries
Range: Alaska, W Canada and U.S.A.

Violet-green Swallow
Size: 4½ in
Diet: Flying insects
Range: Alaska, W North America; Central America

Western Bluebird
Size: 6½–7½ in
Diet: Insects during summer, fruit and seeds during winter
Range: W and S U.S.A., Mexico

Western Scrub-jay
Size: 11 in
Diet: Insects, seeds, nuts, berries
Range: W U.S.A., Mexico

White-breasted Nuthatch
Size: 6 in
Diet: Insects, seeds
Range: S Canada, U.S.A., Mexico

White-crowned Sparrow
Size: 7 in
Diet: Seeds, buds, berries, insects
Range: Alaska, Canada, U.S.A., Mexico

White-throated Sparrow
Size: 6–7 in
Diet: Seeds, insects
Range: Canada, U.S.A.

Yellow-bellied Sapsucker
Size: 8 in
Diet: Insects, fruit, tree sap
Range: S Canada, E U.S.A., Mexico

Yellow-rumped Warbler
Size: 5–6 in
Diet: Insects, berries
Range: North America, Central America

Glossary

abrasion Wear and tear on feathers, often removing paler spots and fringes and fading darker colors.

albinism A lack of pigment. True albinos are white with pink eyes, but most "white" birds are partial albinos, or albinistic, with patches of white and normal eye colors.

axillaries The feathers under the base of the wing, in the "wingpit." Also known as axillars.

band A metal band placed around a bird's leg, with an individual number; when the bird is caught or found dead, its movements can be traced. Also known as a ring.

beak Synonymous with bill; the two jaws and their horny covering.

bird of prey Usually refers to daytime birds of prey, including eagles, vultures, hawks, falcons, harriers, and kites; may be used to include owls. Also called "raptors," or raptorial birds.

brood A set of young birds hatched from one clutch of eggs.

call note A vocalization, usually characteristic of the species, made to maintain contact, warn of danger, or for other specific purposes.

cap A patch of color on the top of a bird's head, usually on the feathers of the forehead and crown.

carpal joint The bend of the wing, at the "wrist."

chick A young bird before it is able to fly.

clutch A set of eggs laid and incubated together in the nest; some species have several clutches during one breeding season, others ("single brooded") have only one.

colony A group of nests close together, often on the ground or in trees.

color ring or band A plastic or metal band placed on a bird's leg; a combination of colors or numbers on the band allow individual recognition without having to capture the bird.

corvid A bird of the crow family or corvidae.

courtship Usually ritualized behavior—male and female together forming a pair bond before breeding.

cryptic Describes coloration that gives a bird camouflage or makes it harder to see.

dawn chorus The loud chorus of bird song heard in spring from just before dawn.

display A form of ritualized behavior with a specific function, for example in courtship, or in distracting potential predators.

distribution The geographical range of a species, often split into breeding range, wintering range, and areas in which it may be seen on migration.

drake A male duck (females are then "ducks").

drumming The sound made in spring by a woodpecker vibrating its bill against a branch; also made by a snipe diving through the air with outer tail feathers extended and vibrating.

dusting "Bathing" in loose, dry sand, dust, or soil to help remove parasites from feathers.

eclipse A dull plumage worn by male ducks and geese in summer.

extinct Describes a species no longer living anywhere on Earth. If a species has disappeared from a country or region, but is still found elsewhere, it is properly described as having been "extirpated" from that area.

fall A sudden large arrival of migrant birds, especially when caused by bad weather on the coast.

feral Describes a bird or species that has escaped from captivity to live wild.

field "In the field" means "in the wild" or out of doors (as opposed to being captive, or held "in the hand").

field guide An identification guide to birds as they are seen wild and free.

fledgling A young bird that has just learned to fly and has its first covering of feathers.

flock A group of birds behaving in some sort of unison. Tight flocks (e.g., starlings in flight) are obvious, but loose, feeding flocks of birds in woodland may be less so.

game bird Bird commonly shot for sport—usually used to describe one of the pheasant, partridge, grouse, or quail families. Other birds include ducks and geese ("waterfowl").

genus A category in classification, above species, indicating close relationships. Appears as the first word in a two- or three-word scientific name. Plural is "genera."

gorget Band of color or pattern, such as streaks, around the bird's upper breast.

habitat The environment that a species requires for survival. Its characteristics include shelter, water, food, feeding areas, nest sites, and roosting sites. More loosely described in such terms as "lowland heath" or "deciduous woodland;" also used for particular times of year or types of behavior, e.g., muddy estuary, open sea, ploughed fields.

hen A female bird.

immature Describes a bird not yet old enough to breed or have full adult plumage colors.

incubation Maintenance of proper temperature of the egg to allow development of the embryo.

juvenile The young bird in its first full plumage. Also known as juvenal in the U.S.

loafing Sitting or standing, often in groups, apparently doing little or nothing. Gulls, for example, "loaf" for hours at a time.

mandible The jaw and its horny sheath; upper and lower mandibles together form the beak or bill.

measurements The size of a bird is usually indicated by the length from bill tip to tail tip on a bird laid out on a flat surface. In reality, the "size" depends as much on shape and bulk as on length.

migration A regular, seasonal movement of birds from one region or continent to another, between alternate areas occupied at different times of year.

molt The replacement of a bird's feathers, in a regular sequence characteristic of each species. There may be a complete molt or a partial molt depending on the season.

nest A receptacle built to take a clutch of eggs and, in many species, the young birds before they are able to fly; eggs may also be laid on a bare ledge or on the ground, with no nest structure being made.

nocturnal Active at night.

numbers Bird populations vary hugely from season to season, so are best described in terms of a particular measure that is easily repeated, usually "breeding pairs." In the case of large, more easily counted birds, such as ducks and geese, the measure is the total number of individuals at a certain season.

ornithology The study of birds. Usually refers to scientific study of biology and ecology, while the hobby of watching birds is known simply as bird-watching or birding.

passage migrant A species or bird seen in some intermediate area during its migration from summer to winter quarters (or vice versa).

passerine A "perching bird."

plumage A covering of feathers; also often used to describe the overall colors and patterns of the feathers, defining a bird's appearance according to age, sex, and season.

preening Care of the feathers, especially using the bill to "zip" the structures back into place.

race A recognizable geographical group, or subspecies, within a species. Often there is no obvious border between groups, which blend (in a "cline") from one

extreme to another. There may be more distinctive differences between isolated areas, such as islands, in which case the decision whether there are races, or separate species, can be difficult.

rarity An individual bird in an area where it is not normally seen, or is seen in only very small numbers. A species with a small world population is "rare."

roost To sleep; also the area where birds sleep.

seabird A species that comes to land to nest, but otherwise lives at sea and is not normally seen inland.

shorebird A plover, sandpiper, or related species; outside of North America, usually called a "wader." Neither word is entirely satisfactory.

soaring Flight, often at a high level, in which the wings are held almost still, using air currents for lift and propulsion.

song A vocalization with a specific purpose and usually distinctive for each species. In particular, advertising the presence of a bird on its territory.

species A group, or groups, of individuals that can produce fertile young. Different species rarely interbreed naturally; if they do so, infertile hybrid offspring are produced.

territory An area defended for exclusive use by an individual bird or a family. Both breeding and winter-feeding territories may be defended.

waterfowl Ducks, geese, and swans. Also known as wildfowl.

INDEX

A

Acanthorhynchus
     superciliosus 161
Accipiter nisus 52
Accipiter striatus 46, 53
Acridotheres tristis 83
Aegithalos caudatus 27
Aegolius acadicus 58
Aethopyga siparaja 171
African Palm Swift 105
African Paradise
     Flycatcher 139
African Yellow White-eye 163
Alcedo atthis 66
Amblyospiza albifrons 131
American Goldfinch 38
American Kestrel 51
American Redstart 134, 180
American Robin 22
Amethyst Starling 91
Anas platyrhynchos 4, 49
Anthochaera paradoxa 158
Anthreptes platurus 5, 164
Apus apus 104
Archilochus colubris 9, 14
Ardea cinerea 47, 48
Australian Magpie 118

B

Baeolophus bicolor 28
Barn Swallow 94, 110
Barred Antshrike 87
Batis molitor 137
Belted Kingfisher 67
Black-billed Magpie 107
Blackbird 21
Blackcap 175
Black-capped Chickadee 23
Bleating Camaroptera 172
Blue-crowned Motmot 9, 19
Blue-faced Honeyeater 156
Blue Flycatcher 138
Blue-gray Gnatcatcher 140
Blue Jay 8, 45
Blue-throated Barbet 71
Blue Tit 26

Bohemian Waxwing 75
Bombycilla garrulus 75
Boobook Owl 59
Brown Thrasher 79
Bubo virginianus 57

C

Cacatua galerita 62
Camaroptera brachyura 172
Cardinal Honeyeater 134, 159
Cardinalis cardinalis 32
Cardinal Woodpecker 46, 72
Carduelis cabaret 40
Carduelis carduelis 39
Carduelis chloris 37
Carduelis tristis 38
Carpodacus mexicanus 41
Chaetura pelagica 103
Chaffinch 36
Chestnut-eared Bulbul 149
Chimney Swift 103
Chin-spot Puff-back
     Flycatcher 137
Chipping Sparrow 123
Chordeiles minor 65
Chrysolampis mosquitus 5, 13
Cinnyricinclus leucogaster 91
Coal Tit 24
Coccothraustes vespertinus 42
Coccyzus americanus 63
Colaptes auratus 18
Colius striatus 69
Columba livia 101
Columba palumbus 100
Columbina passerina 60
Common Barn-owl 94, 97
Common Grackle 89
Common Ground-dove 60
Common Kestrel *see*
     Eurasian Kestrel
Common Koel 77
Common Moorhen 55
Common Mynah 83
Common Nighthawk 65
Common Pheasant 54
Common Tody Flycatcher
     135, 144

Common Waxbill 120, 130
Corvus albus 93
Corvus monedula 4, 108
Crimson Rosella 8, 12
Cyanistes caeruleus 26
Cyanocitta cristata 8, 45
Cypsiurus parvus 105

D

Dacelo novaeguineae 70
Dark-eyed Junco 31
Delichon urbica 4, 111
Dendrocopos major 8, 15
Dendroica coronata 179
Dendropicos fuscescens
     46, 72
Dicaeum cruentatum 169
Dicaeum hirundinaceum 153
Dickcissel 121, 125
Dumetella carolinensis 78
Dunnock 117

E

Eastern Bluebird
     94, 106
Eastern Kingbird 88
Eastern Phoebe 109
Eastern Screech Owl 56
Elminia longicauda 138
Emberiza schoeniclus 126
Entomyzon cyanotis 156
Eolophus roseicapillus 64
Erithacus rubecula 8, 20
Estrilda astrild 120, 130
Eudynamys scolopacea
     77
Eurasian Goldfinch 39
Eurasian Jay 82
Eurasian Kestrel 50
Eurasian Nuthatch 29
Eurasian Scops Owl
     4, 99
Eurasian Swift 104
Eurasian Tawny Owl 98
European Robin 8, 20
European Starling 44
Evening Grosbeak 42

ACKNOWLEDGMENTS

Illustrations Norman Arlott, Hilary Burn, Chris Christoforou, Robert Gillmor, Peter Hayman, Denys Ovenden, David Quinn, Andrew Robinson, Chris Rose, Ken Wood, Michael Woods